ROYAL BRITISH COLUMBIA MUSEUM HANDBOOK

Sea Stars

OF BRITISH COLUMBIA, SOUTHEAST ALASKA AND PUGET SOUND

PHILIP LAMBERT

Drawings by Gretchen Markle
Photographs by Brent Cooke and Philip Lambert

UBCPress · Vancouver · Toronto

Published by UBC Press in collaboration with the Royal British Columbia Museum.

ISBN 0-7748-0825-X
Printed in Canada.

The front-cover photograph shows a Striped Sun Star (*Solaster stimpsoni*) in the low intertidal zone near Sitka, Alaska. On the back cover, Purple Sea Stars (*Pisaster ochraceus*) are clustered below a bed of California Mussels (*Mytilus californianus*).

Canadian Cataloguing in Publication Data

Lambert, Philip, 1945–
Sea stars of British Columbia, Southeast Alaska, and Puget Sound

(Royal British Columbia Museum handbook, ISSN 1188-5114)
Copublished by: Royal British Columbia Museum.
Includes bibliographical references and index.
ISBN 0-7748-0825-X

1. Starfishes – British Columbia. 2. Starfishes – Washington (State) – Puget Sound. 3. Starfishes – Alaska. I. Royal British Columbia Museum. II. Title. III. Series.
QL384.A8L35 2000 593.9'3'091643 C00-911121-2

UBC Press
University of British Columbia
2029 West Mall
Vancouver, BC, Canada
V6T 1Z2

Tel: 604-822-5959
Fax: 604-822-6083
e-mail: info@ubcpress.ubc.ca
Web: www.ubcpress.ubc.ca

SEA STARS OF BRITISH COLUMBIA, SOUTHEAST ALASKA AND PUGET SOUND

CONTENTS

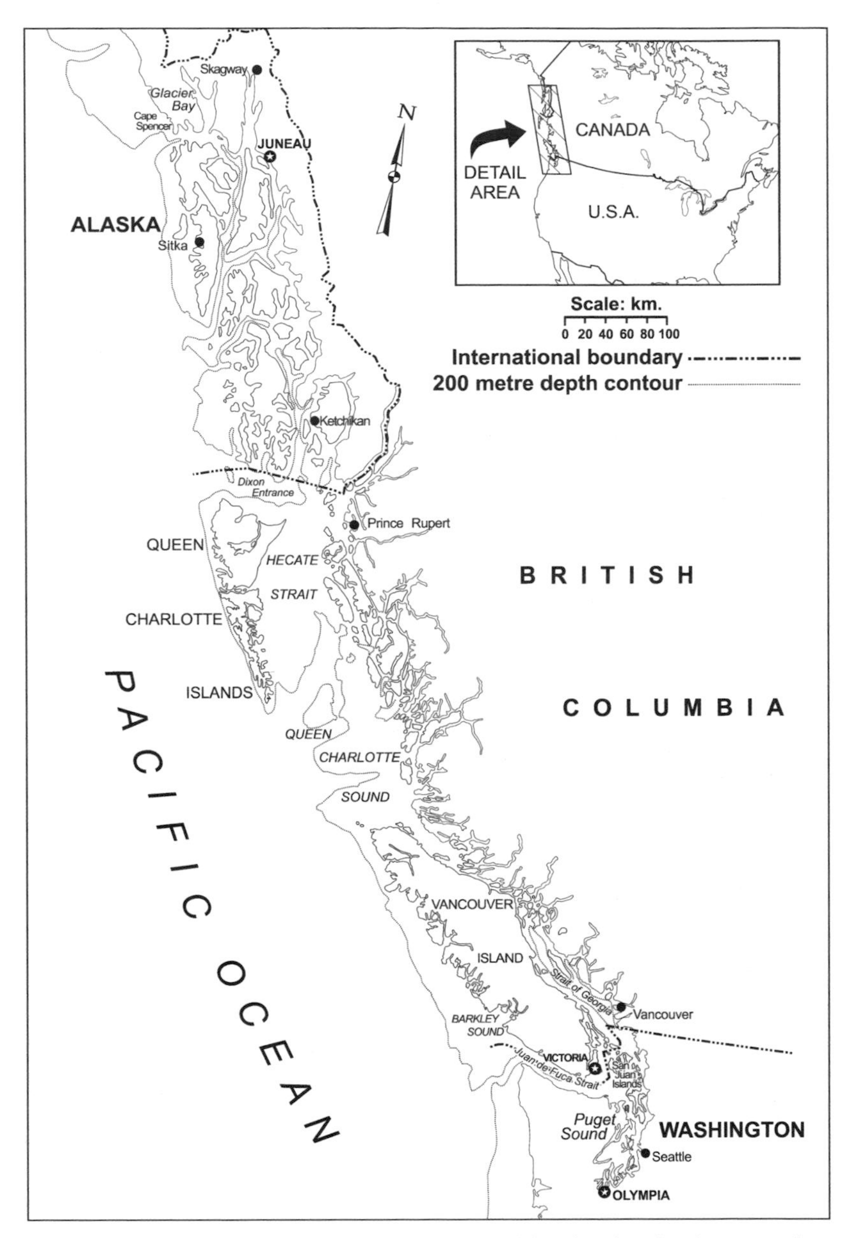

1. The marine waters of the region covered by this book, showing the 200-metre depth contour.

PREFACE

My first handbook on sea stars was published in 1981, covering the species known in British Columbian waters. In this book, I have expanded the geographic area to include a more natural region, from Glacier Bay in southeast Alaska to Puget Sound in northern Washington (see figure 1). The sea-star fauna of this region is the most diverse of all the temperate waters of the world. The great age of the Pacific Basin, and the diversity of habitats along this complex coastline created by scouring glaciers and other natural forces, have encouraged the evolution of many new forms.

In this edition, I describe 43 species and subspecies of sea stars – all that we know of in this region. I also list 26 more whose upper limit is below 200 metres, the depth limit covered by this book, and 14 that occur just north or south of the area shown in figure 1. With further exploration and collection these species might eventually be found within the geographic boundaries of this book.

As with the first edition, my purpose is to bridge the gap between academic and popular publications without sacrificing the details necessary to identify sea stars accurately. I use existing common names and do not attempt to introduce any new ones. Some may criticize this, but to me, a Latin name is more succinct. To aid the reader in understanding and remembering the scientific names of sea stars, I define the Latin and Greek roots.

I obtained the information in this book from published literature and personal communications with marine biologists, and from my own field observations and studies of collections at the Royal British Columbia Museum and elsewhere. Instead of including selected references, I list all that I found to help students and researchers to come up to speed on what is known about each species. A species with a short reference list indicates that we lack knowledge about it. Any constructive criticisms or additional information will be gratefully received so that I may revise and improve future editions.

This edition also gives me the opportunity to add colour photographs. Many were taken in the 1970s and 1980s by Brent Cooke, formerly of the Royal B.C. Museum's Aquatic Zoology Division (as it was called then) and now Director of Public Programs here. I would like to acknowledge his photographic contributions, which are an invaluable resource in our programs to educate the public about the amazing diversity of marine life in the waters around us. Many others who helped with my research and the production of this book are acknowledged on page 178.

INTRODUCTION

What we commonly call sea stars or star fish, scientists place in the Class Asteroidea, part of a larger group of spiny-skinned animals called the Phylum Echinodermata. Besides sea stars, echinoderms include sea urchins, brittle stars, sea cucumbers, feather stars and sea daisies (figure 2). All adult echinoderms are characterized by an internal calcareous skeleton, radial symmetry and a unique network of fluid filled tubes called the water-vascular system.

Sea daisies are a newly discovered group of echinoderms called Concentricycloidea. These tiny animals, no bigger than a dime, were collected in 1000 metres off New Zealand in 1986 and in the Caribbean at 2000 metres. They have the characteristics of echinoderms, but whether they should be placed in a new class or included in Asteroidea remains unclear.

Fossil evidence indicates that the majority of echinoderms first appeared in the Cambrian period (540-500 million years ago), although a form called *Archurua* from the Vendian period (600-540 million years ago) is thought to be the first echinoderm. Relatively soon after, in the Ordovician (500 million years ago), the echinoderms evolved into the five classes we know today, plus several others that have since become extinct. Systematists, who study the evolutionary relationships of living things, have long debated how the classes of echinoderms relate to each other. But there seems to be general agreement now, that sea lilies are the oldest group, followed in order by sea stars, brittle stars, sea cucumbers and sea urchins. See David and Mooi (1998) and Littlewood et al. (1997) for further discussion.

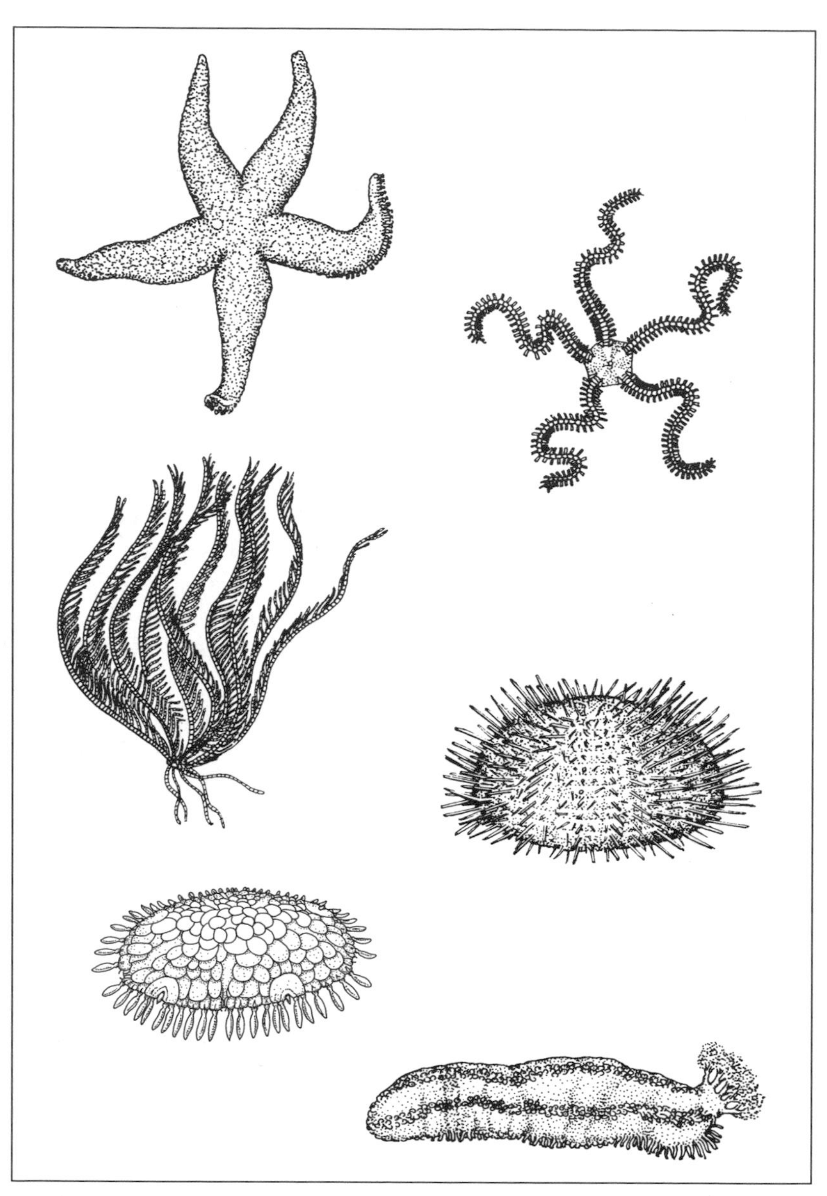

2. The six classes of Phylum Echinodermata (clockwise from top left): Sea Stars (Asteroidea), Brittle Stars (Ophiuroidea), Sea Urchins (Echinoidea), Sea Cucumbers (Holothuroidea), Sea Daisies (Concentricycloidea) and Feather Stars (Crinoidea).

Because of the conspicuous shapes and colours of echinoderms, they have been known since ancient times. The Greek philosopher, Aristotle, recognized three groups of echinoderms but the name echinoderm was not introduced until 1734 when it was first used in reference to sea urchins. For many years the relationship of echinoderms to other groups was in question. The group was, for a time, combined with sea anemones because both have radial symmetry. By 1847, however, biologists realized from studies of internal anatomy and embryology that echinoderms were more complex organisms than sea anemones and should be regarded as a separate phylum.

Sea stars differ from other echinoderms in having five or more open furrows on the undersides of the arms bearing rows of tube feet, digestive glands and gonads radiating into each arm, and respiration by means of dermal gills or papulae. We know of about 2,000 species of sea stars from the intertidal zone to a depth of 7,245 metres, where the deepest known sea star, *Eremicaster vicinus*, was found. Sea stars live in all oceans but none can tolerate fresh water. The greatest number of species occur in the Indo-Pacific, the lowest in the eastern tropical Atlantic. In temperate seas, the North Pacific boasts the greatest diversity of sea stars.

Fisher (1911) notes that ". . . the west coast of North America is more prolific in species and individuals than any other portion of the world." Although that statement is no longer true, he did record 143 species from the shore to a depth of 2,260 fathoms (4,133 metres) between San Diego, California, and Point Barrow, Alaska. Seventy of those species are endemic to the west coast. Nine species occur in the North Pacific and North Atlantic. Several warm periods over the last 15 million years allowed species to migrate through the Arctic Ocean, resulting in faunas of complex origins. Although by morphological criteria they appear to be the same species, DNA analysis might show minor genetic differences.

Eighty three species and subspecies, representing half the world's families of sea stars, occur in the productive waters of Southeast Alaska, British Columbia and Puget Sound, which I refer to throughout this book as "this region". This handbook illustrates and describes 43 that occur between the shore and the outer edge of the continental shelf at a depth of 200 metres.

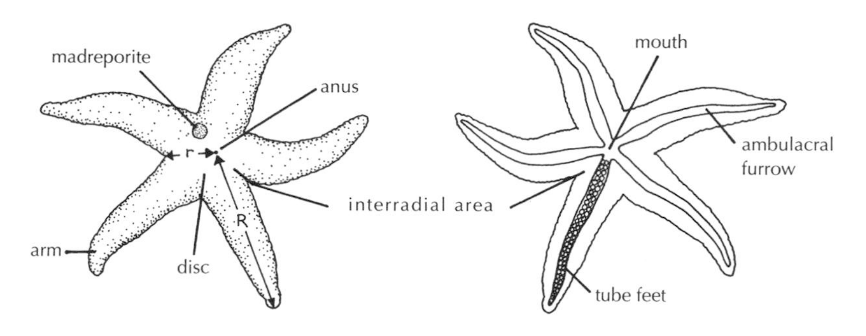

3. The external features of a sea star.

External Anatomy

A typical young sea star is shaped like a five-pointed star; but when mature, some grow additional arms. The genus *Leptasterias* has 5 or 6 arms, *Crossaster* and *Solaster* 7 to 14 arms, and *Pycnopodia* 15 to 24 arms. To describe sea stars accurately and concisely, I use some special terms; they are defined in the Glossary.

Figure 3 illustrates the main external features of a sea star. The top side of the animal is called the aboral side. The central area, or disc, grades into the arms. The triangular region of the disc between the angle of two arms is the interradial area. The distance from the centre of the disc to the edge where two arms meet is the disc radius, designated by "r" (lower case). The arm length or radius of the sea star, measured from the centre of the disc to the tip of the arm is designated by "R" (upper case). The ratio of arm length to disc radius (R : r) is an important measure of body proportions used in the taxonomic descriptions. The madreporite, an obvious coloured patch on the disc, is part of the water-vascular system, which is explained in the next section. On the oral side, closest to the substrate, five or more ambulacral furrows containing the tube feet, or podia, radiate from the central mouth.

To study the surface details of a sea star you will need a hand lens or microscope. The papulae, also called gills or dermal branchiae, are thin-walled, finger-like extensions of the coelom that protrude between skeletal plates to exchange respiratory gasses between the water and the internal fluid. They may occur evenly over the surface, in groups, or in well defined areas, called papu-

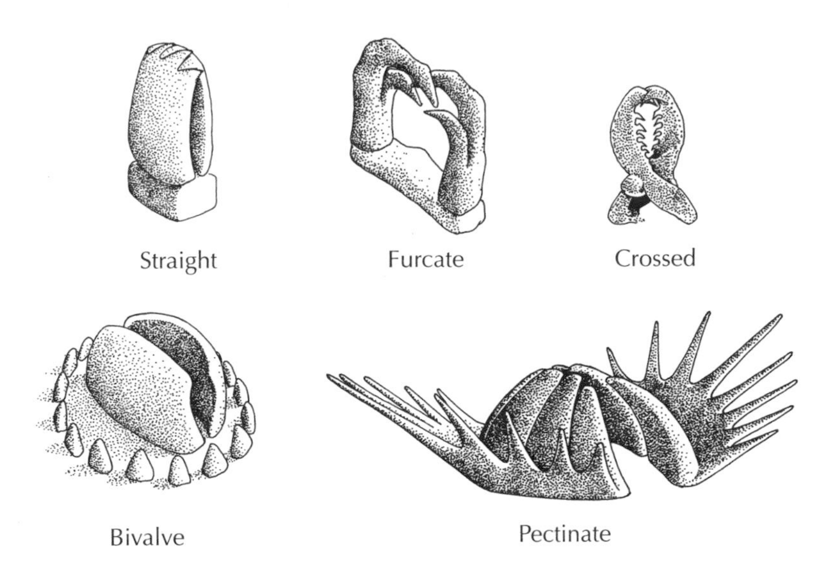

4. Types of pedicellariae.

laria, on the disc or arms. The rest of the surface is occupied by plates bearing spines, granules or tiny pincerlike appendages called pedicellariae; these may play a role in repelling minute organisms that try to settle and grow on the skin. Common types of pedicellariae are illustrated in figure 4.

Aboral skeletal plates, called paxillae, consist of a broad base and an erect column bearing small moveable spinelets (figure 5). The Orders Velatida and Spinulosida have similar structures but the spinelets are extensions of the column and not moveable. These

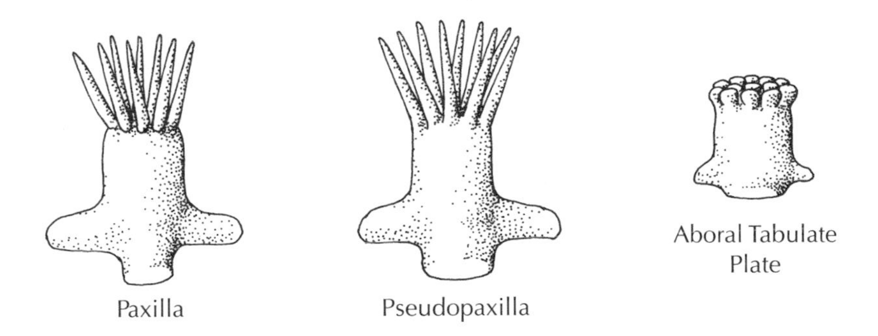

5. Plates of the aboral surface.

are called pseudopaxillae or metapaxillae, but for all practical purposes they look the same. The aboral plates of genera such as *Ceramaster* and *Mediaster* have flat table-like tops and are called tabulate plates. Each species of sea star has a characteristic pattern of plates, spines and granules that range in texture from the flat, smooth surface of *Ceramaster patagonicus* (figure 49) to the spiny surface of *Hippasteria spinosa* (figure 57).

Internal Anatomy

Beneath the skin, the calcareous plates of the internal skeleton vary in shape among species from hexagons to crosses. The linkages between the plates range from a tightly fused meshwork, to a loosely connected netlike pattern. The same plate groupings occur in most sea stars although some may be secondarily reduced or absent. Figure 27 shows the arrangement of the skeletal plates in the arm of a sea star. All spines and other appendages attached to these plates derive their names from the plate; hence, we refer to the superomarginal spines, adambulacral spines, etc.

The skeletal plates of sea stars are held together by a type of tissue called catch connective tissue or mutable collagenous tissue, which is peculiar to all echinoderms. Under nervous control, this tissue can quickly become stiff or soft. For example, the Purple Sea Star (*Pisaster ochraceus*) can be quite stiff when out of water yet when submerged it seems to bend and move quite easily. This is catch connective tissue in action. The soft phase allows normal movement. The stiff phase is activated to anchor the animal in a crevice or firmly attach it to a rock. It is much more energy efficient to use this tissue to maintain a position than to use prolonged muscle contraction. Scientists suggest that the success of the echinoderms may be related to the evolution of this tissue.

The water-vascular system (figure 6) is one of the characteristic features of echinoderms. This hydraulic system operates the tube feet (podia) lining the ambulacral furrow. The madreporite, a porous calcareous plate on the aboral surface of the disc, connects the water-vascular system to the outside environment. Periodically, water is drawn in to replenish fluid in the system. Grooves on the surface of the madreporite snare large particles which are then flushed away. Any particles that do enter the pores are picked up by cilia, hairlike extensions of the cells, and transported away.

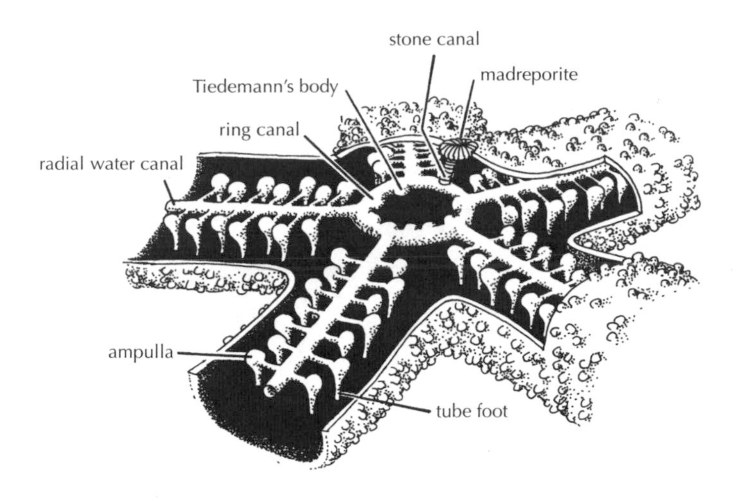

6. The water-vascular system.

Special cells lining the madreporic canal act as the first line of defence against bacteria or other foreign cells that enter with the sea water. From here, the stone canal, a vertical calcareous tube, leads down to the ring canal which circles the mouth. The ring canal in many sea stars has sacs called Tiedemann's bodies containing cells that also capture and digest foreign bacteria that have invaded the system. These bodies also function in producing coelomic fluid, which is mixed with incoming sea water and recirculated. Radial water canals lead off this ring to each arm, giving off lateral branches at regular intervals to supply the tube feet. Each lateral branch has a muscular valve that allows the tube foot to be isolated from the rest of the system. To extend the tube foot, the valve closes, the bulb (ampulla) contracts and fluid pressure extends the tube foot lengthwise. To withdraw the tube foot, the valve opens to release the pressure as the longitudinal muscles contract.

An inconspicuous system of tubes called the haemal system runs roughly parallel to the water-vascular system but also has connections to the gonads. Although the name suggests a blood system, its true function is not clear. The main organs of the sea star in the body cavity (coelom) are bathed in a fluid that performs the function of blood by transporting dissolved nutrients and respiratory gasses. The coelomic fluid contains wandering cells called

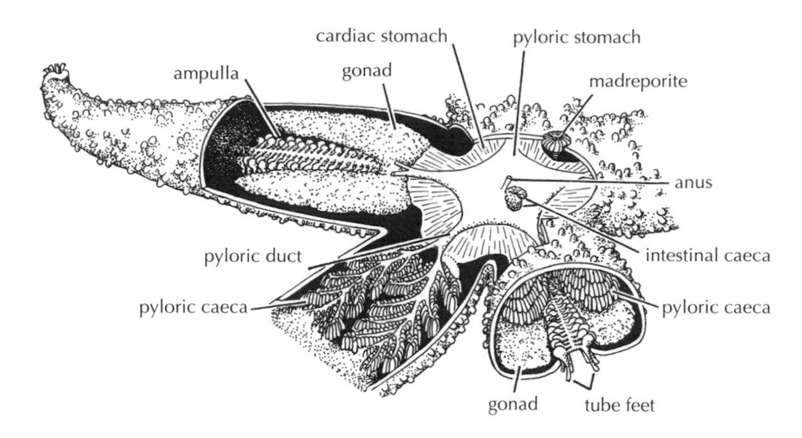

7. The internal organs of a sea star.

coelomocytes, which engulf bacteria and waste products and perhaps assist in healing wounds.

The digestive system (figure 7) occupies a large part of the body cavity. The mouth leads into the cardiac stomach and above this is a smaller pyloric stomach. Two pyloric ducts radiate from the pyloric stomach in each arm, branching immediately into soft, lobed glandular organs called pyloric caeca. Here, the gland cells secrete all the enzymes necessary for digestion. Other cells absorb or finish the digestive process. Pyloric caeca serve as the major food store, akin to our liver, and become enlarged during times of abundant food. From the top of the pyloric stomach a short intestine leads to the anus in the centre of the disc. Members of the families Luidiidae and Astropectinidae lack an intestine or anus; but this is considered a recent specialization rather than a primitive condition like that found in many fossil sea stars. Close to the anus are intestinal caeca, small glands that help compact waste matter. Sea stars have no excretory system for liquid waste but seem to pass it directly through the walls of the tube feet.

The nervous system of a sea star is unusual, because the major portion of it is external and consists of thickenings of the skin specialized to conduct nerve impulses. There is no central brain. The main parts are a nerve ring encircling the mouth and radial nerve cords running the length of each arm in the midline of the ambulacral furrows. Smaller adradial nerves run in the skin beside the tube feet (figure 8). Except for some obvious nerves to internal

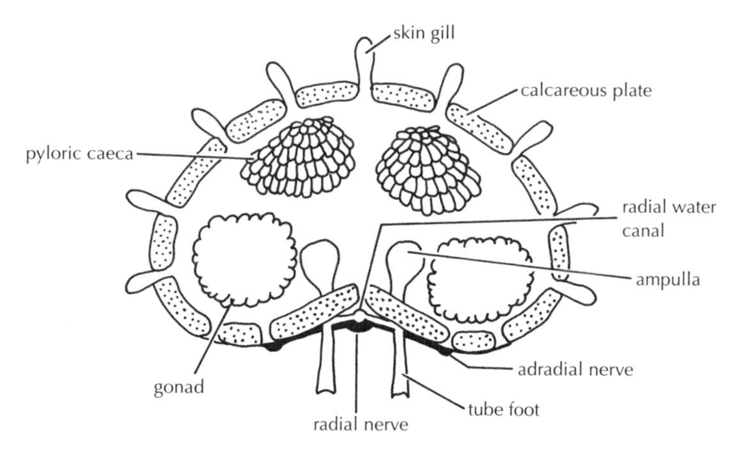

8. Cross section of an arm.

organs, the rest of the nervous system is a diffuse network of nerve cells embedded in the skin. The only conspicuous sense organs are brightly coloured eye-spots (ocelli), one at the tip of each arm. The ocelli are sensitive to light and have a simple lens. In those of *Asterina miniata*, small extensions of the cell membrane called microvilli are sensitive to light; they deteriorate in bright light and are rebuilt during darkness.

Reproduction

Most species of sea stars have separate sexes. Paired gonads in each arm (figures 7 and 8) open to the outside by gonopores, usually located where two arms join. Most sea stars shed eggs and sperm directly into the water and fertilization takes place there. The free-swimming larvae that develop can be one of two forms: planktotrophic larvae have a mouth and feed on other plankton; lecithotrophic larvae have no mouth and survive on a stored yolk supply.

A few species brood their eggs. In these, the gonopores open on the oral side of the disc. Males shed their sperm in the normal way but the female retains the fertilized eggs beneath her arched body, or in some other specialized cavity. The larvae develop here and crawl away as tiny sea stars. The small Six-armed Star (*Leptasterias hexactis*) broods its eggs.

A typical planktotrophic larva goes through several free-swimming stages (figure 9). At first, the entire body is covered with cilia, which beat in unison to provide locomotion. Later, the cilia aggregate into bands and arms develop as the gut forms and the larva begins feeding. This bipinnaria larva is characteristic of sea stars and may last several weeks or months. This is followed by a brachiolaria larva, which adds three more arms with adhesive tips and a sucker at the mouth end. About this time, the brachiolaria larva seeks a suitable type of substrate and attaches itself by its brachiolar arms. While attached, the bilaterally symmetrical larva metamorphoses to a radially symmetrical sea star. The larval digestive tract degenerates and is formed anew in the juvenile sea star, which now appears as a tiny (less than a millimetre in diameter), stubby armed sea star with a few pairs of tube feet. For details of larval development, consult the species accounts in this book or general reviews such as Boolootian 1966, Chia and Walker 1991, Hyman 1955, Kasyanov et al. 1998, Kumé and Dan 1968, and Strathmann 1987.

Regeneration

All sea stars have strong powers of regeneration. Some species use this ability in asexual reproduction by splitting in two across the disc, in a process called fissiparity, each half regenerating into a whole sea star. *Linckia columbiae* of southern California goes a step further and voluntarily drops arms, which grow into complete sea stars. No sea stars in British Columbia waters are known to reproduce by fissiparity, but the regeneration of lost arms is common. As a means of defence, arms may readily detach when a species is handled excessively or attacked by a predator. Species such as *Luidia foliolata*, *Pycnopodia helianthoides* and *Stylasterias forreri* exhibit this behaviour. Detached arms will only grow into a whole sea star if a portion of the central disc is included. In *Leptasterias hexactis* the arm stub is first sealed off, then damaged tissues are cleaned out by wandering cells (phagocytes). Specialized cells from adjacent damaged tissues such as the water-vascular system, the lining of the gut, nerves, and the skin, migrate into the area and begin to form a complete new arm tip just below the temporary seal, followed by accelerated growth of the arm.

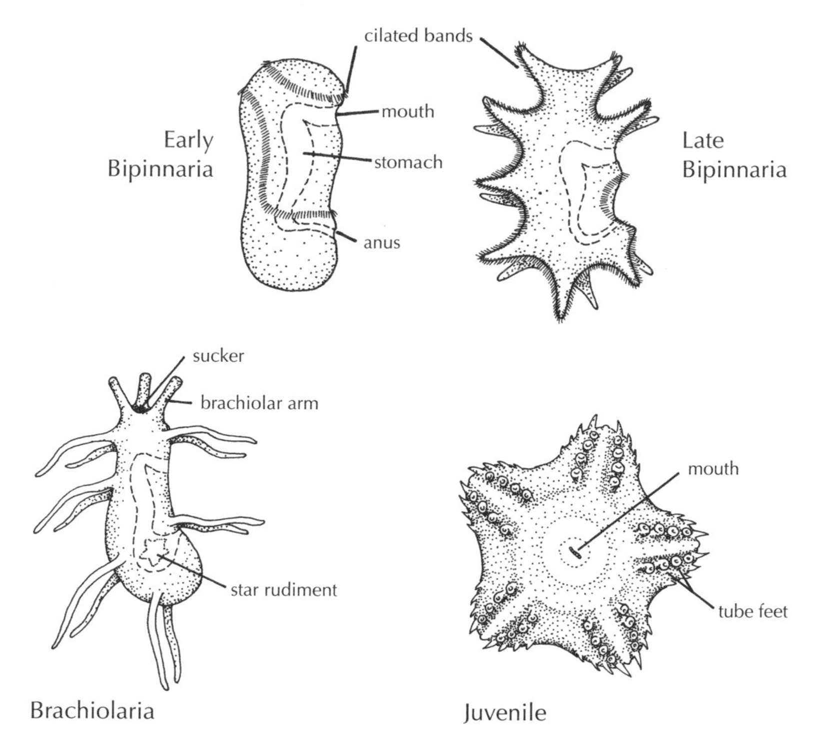

9. The larval stages of a sea star.

Movement

Sea stars move primarily by walking on their tube feet, which may be pointed, as in the Sand Star (*Luidia foliolata*), or possess suckers, as in the Sunflower Star (*Pycnopodia helianthoides*) and most other sea stars. The stepping motion of the tube feet may be combined with a lateral movement of the arm in those species with flexible skeletons. Flexibility is determined by the rigidity of the skeletal plates and the state of the catch connective tissue described earlier. The Purple Sea Star (*Pisaster ochraceus*) has a heavily calcified skeleton, and moves and changes its shape slowly. On the other hand, the Sunflower Star has a flexible body and can move rapidly, because the aboral skeletal plates are only loosely connected to each other.

Feeding

The feeding habits of each species vary depending on its mobility, the type of tube feet it has and its digestive abilities. In temperate waters, most sea stars are predators of attached or buried animals, such as bivalves, sea anemones, sea squirts and barnacles (figure 10). The more mobile species can capture animals like snails, sea urchins or even other sea stars. *Stylasterias forreri*, by an ingenious use of its pedicellariae, can snare some kinds of bottom fish.

Many species from the deep ocean, where prey are scarce, eat mud and digest the organic material. The Mud Star (*Ctenodiscus crispatus*) gets nutrition by consuming organically rich mud. Others are scavengers that eat dead animals, seaweed or detritus.

Less commonly, sea stars obtain nutrition by suspension feeding. When tiny floating food particles come in contact with the mucus covered surface of a sea star, they adhere to the mucus and are wafted by cilia along grooves to the mouth. The Blood Star (*Henricia leviuscula leviuscula*) is probably a suspension feeder.

Experiments have shown that some sea stars can absorb organic molecules, such as amino acids and simple sugars, directly from sea water through their skin. External structures like pedicellariae or the skin itself do not receive nutrients via a blood system and may rely on these dissolved organic substances for nutrition.

Many sea stars feed on large objects by protruding the lower cardiac stomach through the mouth to envelop the prey. Digestive

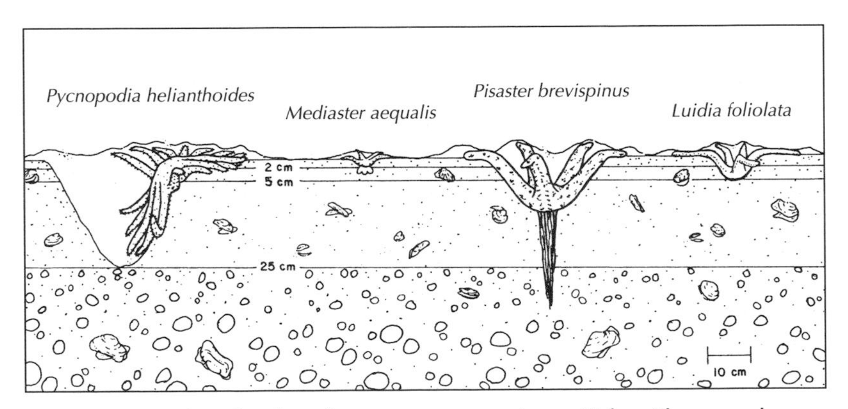

10. Comparative depths of sea-star excavations. (After Sloan and Robinson 1983, p.130.)

enzymes secreted by the pyloric caeca flow to the pyloric stomach then to the cardiac stomach where breakdown of the food begins. Large particles then pass to the pyloric stomach where further digestion takes place, and finally to the pyloric caeca where individual cells take in small particles of food.

Sea stars, exposed at low tide, are usually inactive, so it is unusual to see them feeding. Scuba divers are more likely to observe sea stars during their food-gathering activities: the Morning Sun Star (*Solaster dawsoni*) attacking a Leather Star (*Dermasterias imbricata*), the Rose Star (*Crossaster papposus*) eating a sea slug, the Sunflower Star (*Pycnopodia helianthoides*) swallowing a Purple Sea Urchin or perhaps a Cushion Star (*Pteraster tesselatus*) nibbling at a sponge. A diver can easily document a sea star's prey items by turning it over, because it takes several hours to digest its prey in the protruding stomach.

Encounters between predator and prey often produce predictable behaviour patterns. For example, the abalone reacts to a Sunflower Star by speeding away and swivelling its shell violently in order to shake off the suckers of the attacker. Many snails and limpets can move quickly to avoid a predator. Purple Sea Urchins can also speed away, but they will simultaneously flatten their spines to expose hundreds of pedicellariae.

Each species account gives details of feeding behaviours. For more information, consult a review such as Jangoux 1982.

Parasites and Commensals

Parasites of sea stars in British Columbia waters are poorly documented and could be a productive area of research. Known parasites are indicated in the species accounts. Recently, an introduced species of parasitic ciliate has been discovered in the Purple Sea Star in British Columbia. *Orchitophrya stellarum*, endemic to the North Atlantic, infests the male gonad causing castration. Other internal parasites include ciliated protozoa, flatworms, snails and barnacles. The parasitic barnacles are highly modified, having lost all appendages and consisting almost entirely of reproductive organs. *Dendrogaster arbusculus* has been found in *Hippasteria californica*, *Dendrogaster punctata* in *Poraniopsis inflatus inflatus* and *Dendrogaster fisheri* in *Pedicellaster magister*. Externally, sea stars are known to harbour parasitic crustaceans such as copepods and caprellid amphipods among their spines. Parasitic snails insert their sucking mouth parts through the body wall of sea stars and suck out the internal fluids.

An animal that gains benefit from a host animal without benefiting or harming the host is called a commensal. A group of polychaete worms, called scale worms, are well-known examples of commensals on sea stars. Three species of *Arctonöe* are commonly found among the aboral spines of sea stars but more often in the ambulacral furrow. The scale worm eats scraps of food left over from the sea star's meal. Scale worms can recognize their normal host species either by contact or by chemical substances given off by the host. Specific examples of commensal scale worms are described in the species accounts.

Predators

Sea stars have few predators. In the intertidal zone, gulls will eat small sea stars. Subtidally, *Solaster dawsoni* and *Pycnopodia helianthoides* will attack other sea stars, and in Alaska, the Alaska King Crab and sea otters may attack sea stars, but most adult sea stars have few enemies. Their numbers tend to be controlled at the larval and juvenile stages by predators, and later in life by the food supply.

Species Names

A species is a set of similar individuals capable of breeding successfully with one another in nature and reproductively isolated from other species. Taxonomists usually differentiate species on the basis of anatomical structures, and may only later test these hypotheses using other data.

All species are designated by a scientific name consisting of two words, based on Latin or Greek roots, which are recognized by scientists throughout the world. Common names, on the other hand, are not regulated or governed by strict rules. One species may have several common names depending on local conventions; conversely, one common name may actually include several species.

The scientific name (binomen) consists of two parts; e.g., *Hippasteria spinosa*. The first word is the genus; the second is the species. The species name is not capitalized and is never used by itself. The name of the person who first described the species is customarily placed after the species name, followed by the year the description was published; hence, *Hippasteria spinosa* Verrill, 1909.

Categories broader than the species level, such as genus or family, attempt to group species with a common ancestor. As more species are discovered and described, these broader categories may be modified to accommodate new species. A change in the genus is a common type of revision. This is indicated by brackets around the author and date; e.g., *Pisaster ochraceus* (Brandt, 1835). In this example, Brandt used the generic name, *Asterias*, in his original description. To find out the complete history of name changes consult a scientific publication that indicates the most recent name with a list of all the names (synonyms) previously used for that species.

Sometimes several biologists independently give different names to the same species. In this case, the name published first takes priority, and later ones become synonyms. In this book, I have mentioned only those synonyms that might be encountered in other publications on the marine life of the west coast of North America. Complete synonymies are contained in the publications listed at the back of this book.

A subspecies is a physically distinct population within a species; it inhabits a separate geographic area or depth range within the range of the species. A subspecies is designated by a three-part name (trinomen) consisting of the genus, species and subspecies; e.g., *Henricia leviuscula annectens*. The subspecies that is closest in form to the original species description usually takes the same name as the species (e.g., *Henricia aspera aspera*), but revisions like this may not be made until years later after further study of all available museum specimens. In the meantime, the species may not be fully subdivided into subspecies. Fisher and his contemporaries used the concept of formae to designate different varieties or forms of a species that might occur together in the same region. According to the third edition of the *International Code of Zoological Nomenclature* the terms formae and variety are no longer used but should either be raised to subspecies or reduced to an infrasubspecific name.

Collecting

Large, colourful sea stars are relatively easy to find and are often the most obvious intertidal animals. Of the species in this book, 18 can be collected by hand in the intertidal zone, 32 species are accessible within normal scuba depth, and 10 usually require a dredge or trawl. Sea stars sometimes crawl into crab traps or cling to baited hooks.

To help conserve marine life, identify the specimen in the field and be sure to return it to the place where you found it, or keep it alive in an aquarium and return it to its natural habitat once you have identified it. Or better yet, leave it where it is and take a photograph.

Sea stars and other marine animals should not be collected for souvenirs. For that reason, I have not included recipes for preserving specimens. Colour photographs of the live specimen are preferable. If you have a legitimate purpose and a scientific collecting permit, you can obtain information on preserving specimens from various technical sources. The common methods involve corrosive, poisonous and carcinogenic chemicals that require proper lab facilities. Besides being dangerous, none of these methods preserves the living colour.

General References

I used several major publications that deal with aspects of all the species, so rather than cite them repeatedly throughout the text, I highlight them here. Other more specific references appear at the end of each species account.

For general information about the life and habits of sea stars Ricketts et al. 1985 is still one of the most readable accounts of west coast marine life and a good reference to have on your shelf. MacGinitie and MacGinitie 1968, first published in 1949, has a chapter full of behavioural and ecological observations on echinoderms of the west coast. Similarly, Kozloff 1983 describes the intertidal ecology of most of our common intertidal sea stars. For the Alaska region, Barr and Barr 1983 covers common intertidal and subtidal life with species accounts and colour photos. O'Clair and O'Clair 1998 is an excellent book on the intertidal life of Southeast Alaska, packed full of details about common intertidal species of sea stars and other groups. For beautiful colour photos of the common sea stars of the west coast, consult Gotshall 1994. Morris et al. 1980 and Smith and Carlton 1975 cover the intertidal invertebrates of California. The latter is an identification guide with illustrations and keys.

The most important reference used in the production of this book is W.K. Fisher's three-part monograph on the distribution and taxonomy of sea stars of the North Pacific (Fisher 1911, 1928 and 1930). Fisher describes about a dozen new species and provides extremely detailed accounts of every species in this book. I can appreciate the time and effort that went into researching and writing more than a thousand pages (with 296 plates). For the Bering Sea and Alaska, D'Yakonov 1950 provides descriptions and distribution data. Fisher's contemporary, A.E. Verrill was an expert on Atlantic species and had some taxonomic disagreements with Fisher, but his work (Verrill 1914) is nevertheless an important taxonomic reference. Libbie Hyman devoted an entire volume (1955) of her six-part series to the echinoderms, providing detailed anatomical drawings and ecological information for many west coast sea stars as well as others around the world. If you are interested in the internal workings of sea stars Lawrence 1987 provides a detailed review of the world's literature on the functional biology of echinoderms. The microanatomy of asteroids is reviewed in Chia and Koss 1994.

Alton 1966 surveyed the sea star fauna off the mouth of the Columbia River and added a number of range extensions, especially for deep water species. Austin 1985 provides a detailed checklist of marine invertebrates for the west coast with distribution data for all the known sea stars. Grainger 1966 covers the Arctic species and Maluf 1988 documents many of our northern species that occur in California and south. I extended the ranges of several species while preparing the first and second editions of this book (Lambert 1978a, 1978b, 1981a and 1999). Kozloff 1987 is the reference most of us turn to for identifying invertebrates in the B.C.-Washington area. It contains identification keys for all the major shallow-water taxa, including sea stars.

I consulted the following papers for data on diet: Carey 1972, which provides diet information for subtidal and deep-water species; and Mauzey et al. 1968, an extremely valuable resource that records the feeding behaviour and diets of 18 species of shallow-water asteroids in the Puget Sound region, collected during

300 hours of scuba diving; Hopkins and Crozier 1966 summarizes the data on 18 species of asteroids from southern California, 10 of which occur in our northern waters; and Jangoux 1982 reviews echinoderm feeding studies worldwide. The best single source for data on reproduction of west-coast Asteroidea is Strathmann 1987.

KEY TO FAMILIES

This key is for the families of shallow-water (0-200 metres) sea stars of British Columbia, Southeast Alaska and Puget Sound. It is designed for fresh specimens or those that have been preserved in alcohol. In most cases I have indicated where dried specimens might deviate from the key. Use the key only for specimens collected between Southeast Alaska and Puget Sound above a depth of 200 metres. It consists of pairs of statements. Starting at number 1, choose the statement that best describes your specimen. Your choice will lead to another pair of statements and eventually to the name of a family and a page number. Within each family in the text, the species are arranged alphabetically by genus and species.

1a. Five arms; tube feet pointed (figure 11); white, yellow or grey colour in life; aboral plates paxilliform. 2
1b. Five or more arms; tube feet with suckers (figure 12); aboral plates may or may not be paxilliform; includes most sea stars 4

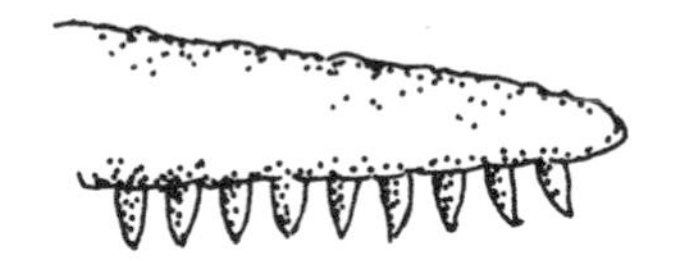

11. Pointed tube feet.

12. Suckered tube feet.

2a. Fleshy cone the in centre of the disc (figure 13); this may not be obvious in a dried specimen .... Ctenodiscidae (p. 44)

2b. No fleshy cone 3

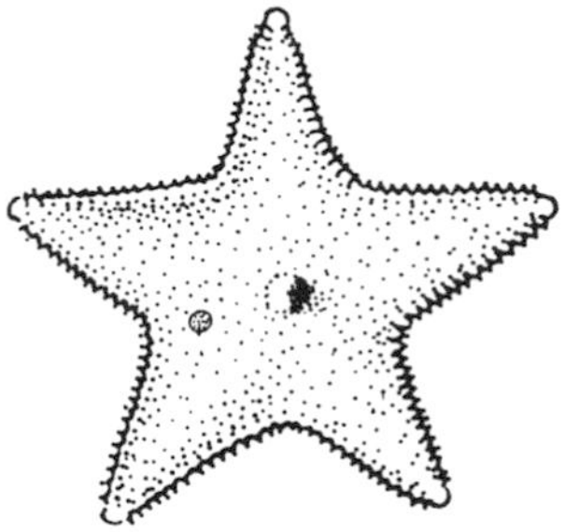

13. *Ctenodiscus.*

14. *Leptychaster.*

3a. Marginal plates define the perimeter of the body when viewed from above (figure 14)....... Astropectinidae (p. 37)

3b. Marginal plates not obvious from above; arms long and rectangular (strap shaped) in cross section (figure 15) .... Luidiidae (p. 34)

15. *Luidia.*

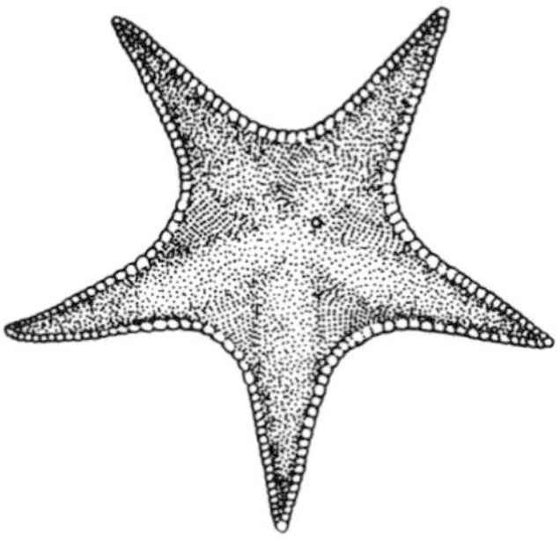

16. *Mediaster.*

4a. Distinct marginal plates, with or without spines, define the perimeter of the body; the aboral surface has tabulate plates or spines (figure 16) 5

4b. Marginal plates not obvious or only on the sides of the arms (but in *Dermasterias imbricata*, the marginal plates may show through the leathery skin of dried specimens) 6

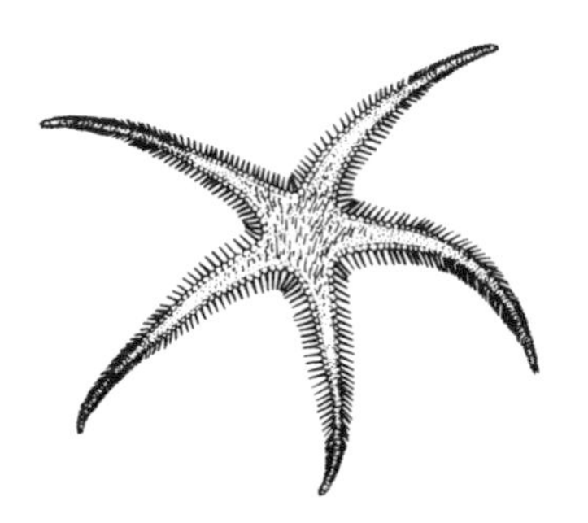
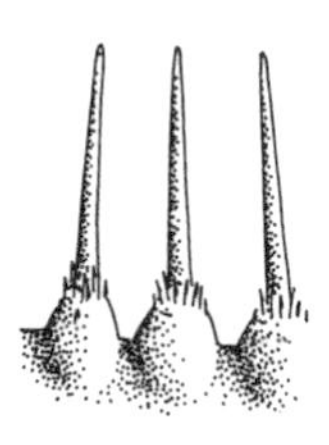

17. *Cheiraster* [formerly *Luidiaster*] and a close-up of its marginal spines.

5a. Marginal plates with long, sharp spines (figure 17, right); arms long, flat and tapering; pectinate pedicellariae (figures 4 and 39); superomarginals and inferomarginals alternating . . . . Benthopectinidae (p. 47)

5b. Marginal plates without long spines (except *Hippasteria spinosa* with stout spines); superomarginals more or less in line with inferomarginals; body pentagonal to star-shaped (figure 18) . . . . Goniasteridae (p. 57)

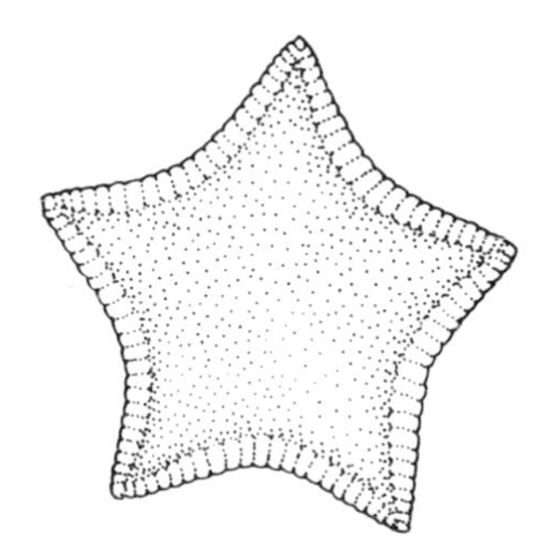

18. *Ceramaster* (left) and *Pseudarchaster* (right).

6a. Has a supradorsal membrane supported by tips of aboral spines; a pore, or osculum, in the centre of the disc. In a dry specimen the osculum may be closed, but the position is indicated by a cluster of spines near the centre (figure 19) . . . . Pterasteridae (p. 91)

6b. No supradorsal membrane or osculum 7

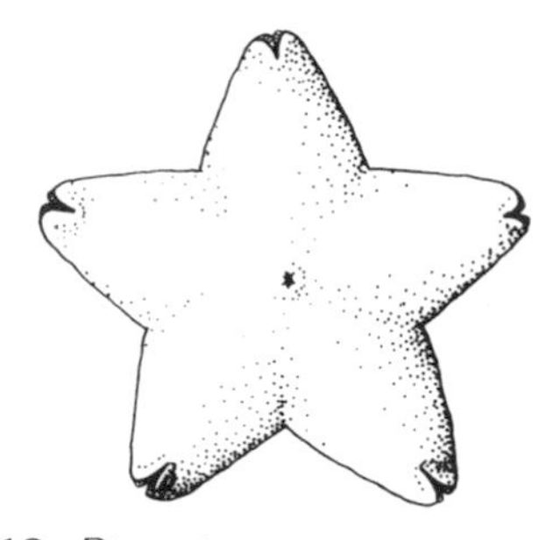

19. *Pteraster.*

20. *Dermasterias imbricata.*

7a. Aboral surface smooth except for occasional calcareous granules (figure 20) . . . . Asteropseidae (p. 71)

7b. Aboral surface not smooth; spines, tabulate plates or paxillae present 8

8a. Aboral surface with paxillae or tabulate plates, or a meshlike pattern bearing pseudopaxillae with clusters of spinelets or single sharp spines; no pedicellariae; two rows of tube feet (figure 21) 9

8b. No paxillae or tabulate plates; aboral surface with sharp or blunt spines usually surrounded by pedicellariae; pedicellariae always present; adoral carina (figure 112) present; four rows of tube feet (figure 22) . . . . Asteriidae (p. 112)

9a. Small tabulate plates interspersed with larger crescent-shaped ones; arms broad and triangular; perimeter of body sharp, marginals inconspicuous (figure 23) . . . . . Asterinidae (p. 50)

9b. Not as in (a) 10

10a. Has more than five arms; or if only five arms, the aboral surface has a coarse meshwork bearing single large, spiny pseudopaxillae on the junctions (figure 24) . . . . Solasteridae (p. 75)

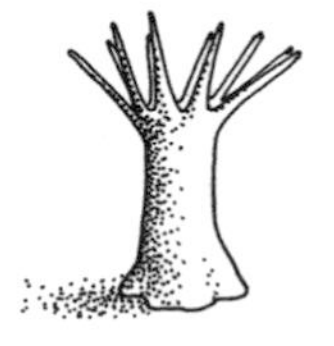

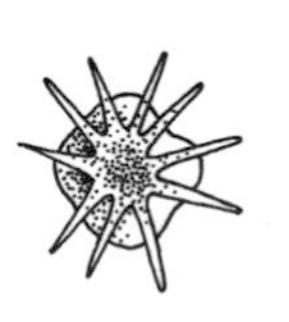

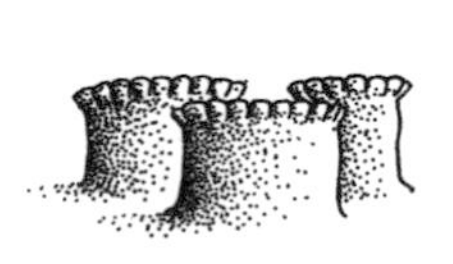

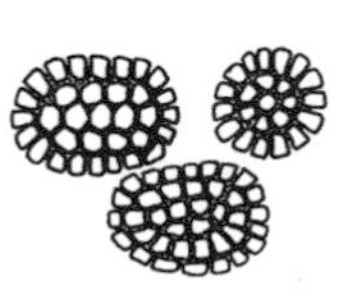

21. Side and top views of paxillae (left) and tabulate plates (right).

22. *Pisaster.*

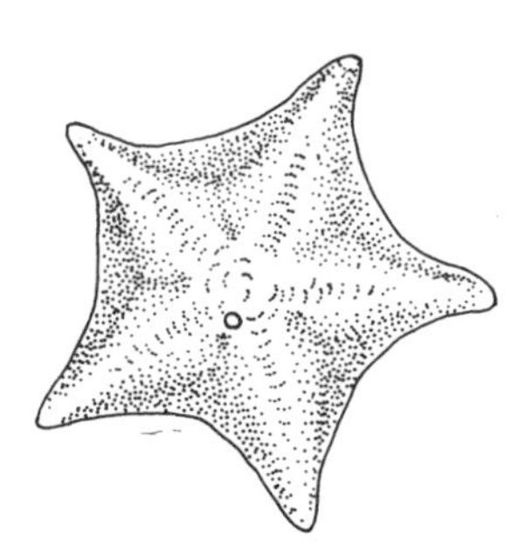

23. *Asterina miniata.*

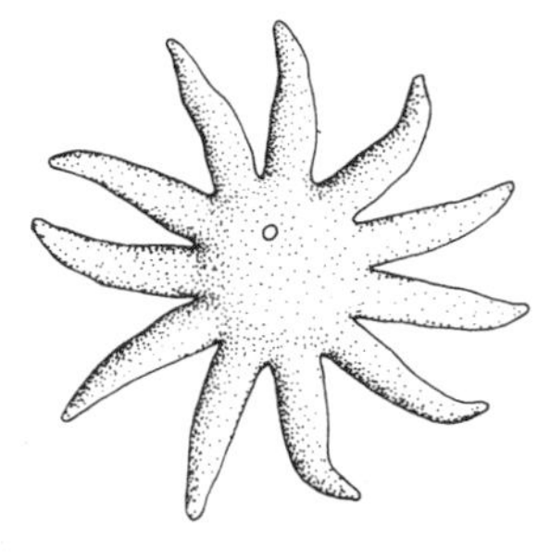

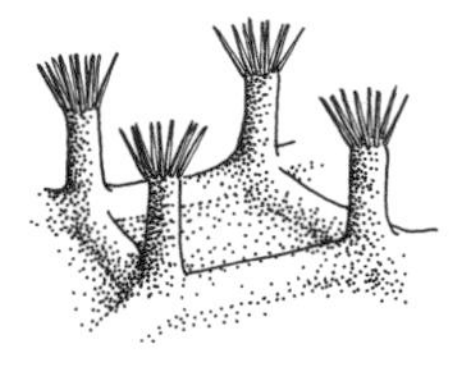

24. *Solaster* and its paxillae-bearing meshwork.

10b. Five arms and not as in (a) 11

11a. Aboral surface of tightly packed ridges with spines or a meshwork bearing clusters of spinelets (figure 25) . . . . Echinasteridae (p. 99)

11b. Aboral surface with an open meshwork with a single sharp spine at each junction (figure 26) . . . . Poraniidae (p. 54)

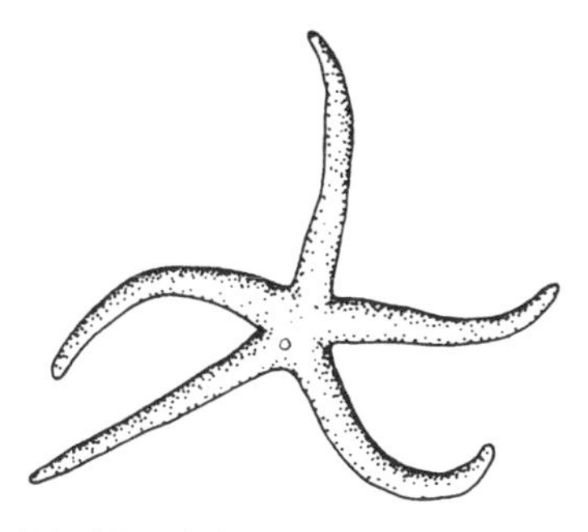

25. *Henricia.*

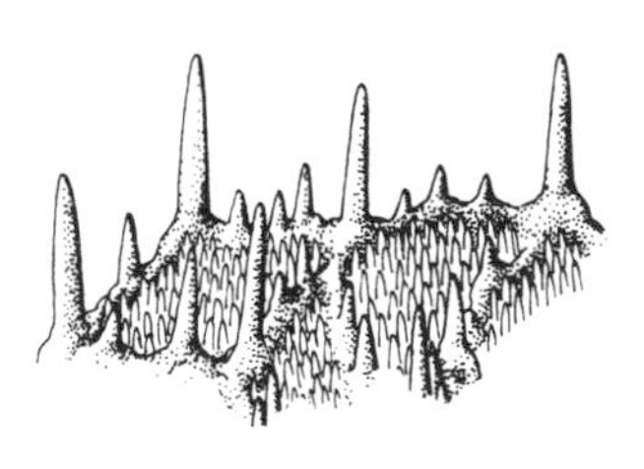

26. Single sharp spines of *Poraniopsis.*

CHECKLIST

This checklist contains all the species known from this region and immediately adjacent areas. Species in bold occur in less than 200 metres and are described in detail in this book. Species that occur just to the north of this region are marked with a superscript "N" and those just to the south with "S"; species marked with "ND" or "SD" occur below 200 metres in this region but shallower to the north or south. The numbers on the right represent the depth range in metres and apply to the species' entire geographic range. Within each family, species are listed alphabetically. (Classification after Clark 1989, 1993 and 1996; Clark and Downey 1992)

Phylum Echinodermata

Class Asteroidea

Order Paxillosida

Family Luidiidae

Luidia foliolata Grube, 1866 — 4–613

Family Astropectinidae

Astropecten verrilli de Loriol, 1899 [S] — 18–445
Dipsacaster anoplus Fisher, 1910 — 362–2202
Dipsacaster borealis Fisher, 1910 — 221–732
Leptychaster anomalus Fisher, 1906 — 59–1258
Leptychaster arcticus (Sars, 1851) — 40–1261
Leptychaster inermis (Ludwig, 1905) — 1200–2000
Leptychaster pacificus Fisher, 1906 — 10–435

Psilaster pectinatus (Fisher, 1905)	1500–3000
Thrissacanthias pencillatus (Fisher, 1905) [SD]	55–1500

Family Porcellanasteridae

Eremicaster crassus (Sladen, 1883)	1570–6330
Eremicaster pacificus (Ludwig, 1905)	1570–4090

Family Ctenodiscidae

Ctenodiscus crispatus (Retzius, 1805)	10–1890

Order Notomyotida

Family Benthopectinidae

Benthopecten claviger claviger Fisher, 1910	1647–1950
Benthopecten mutabilis Fisher, 1910	2870
Cheiraster dawsoni (Verrill, 1880)	73–384
Nearchaster aciculosus (Fisher, 1910)	549–1492
Nearchaster variabilis (Fisher, 1910) [ND]	198–642
Pectinaster agassizi evoplus (Fisher, 1910)	1801–2200

Order Valvatida

Family Asterinidae

Asterina miniata (Brandt, 1835)	0–302

Family Poraniidae

Poraniopsis inflatus inflatus (Fisher, 1906)	11–366

Family Goniasteridae

Ceramaster arcticus (Verrill, 1909)	0–186
Ceramaster clarki Fisher, 1910	611–1098
Ceramaster japonicus (Sladen, 1889) [ND]	195–1438
Ceramaster patagonicus (Sladen, 1889)	10–245
Cryptopeltaster lepidonotus Fisher, 1905	360–1244
Gephyreaster swifti (Fisher, 1905)	11–344
Hippasteria californica Fisher, 1905	300–2000
Hippasteria spinosa Verrill, 1909	10–512
Mediaster aequalis Stimpson, 1857	0–293
Pseudarchaster alascensis Fisher, 1905	92–1947
Pseudarchaster dissonus Fisher, 1910	1435–1947
Pseudarchaster parelii (Duben and Koren, 1846) [ND]	15–2500

Family Asteropseidae

Dermasterias imbricata (Grube, 1857)	0–91

Order Velatida

Family Solasteridae

Crossaster papposus (Linnaeus, 1767)	0–1200
Lophaster furcilliger Fisher, 1905	350–2010
Lophaster furcilliger vexator Fisher, 1910	21–670
Solaster borealis (Fisher, 1906)	301–1922
Solaster dawsoni Verrill, 1880	0–420
Solaster endeca (Linnaeus, 1771)	0–475
Solaster paxillatus Sladen, 1889	11–640
Solaster stimpsoni Verrill, 1880	0–60

Family Pterasteridae

Diplopteraster multipes (Sars, 1865)	57–1171
Hymenaster koehleri Fisher, 1911 [N]	3240
Hymenaster perissonotus Fisher, 1910	412–3241
Hymenaster quadrispinosus Fisher, 1905	1098–3241
Pteraster jordani Fisher, 1905 [S]	128–1801
Pteraster marsippus Fisher, 1910 [N]	95–642
Pteraster militaris (O.F. Müller, 1776)	10–1100
Pteraster tesselatus Ives, 1888	6–436

Family Korethrasteridae

Peribolaster biserialis Fisher, 1905 [N]	104–805

Order Spinulosida

Family Echinasteridae

Henricia aspera aspera Fisher, 1906	6–904
Henricia asthenactis Fisher 1910	91–1250
Henricia leviuscula annectens Fisher, 1910	10–228
Henricia leviuscula leviuscula (Stimpson, 1857)	0–400
Henricia leviuscula spiculifera (H.L. Clark, 1901)	9–680
Henricia longispina longispina Fisher, 1910	28–512
Henricia sanguinolenta (O.F. Müller, 1776)	15–518

Order Forcipulatida

Family Zoroasteridae

Myxoderma sacculatum Fisher, 1905	1007–1546
Zoroaster evermanni Fisher, 1905	395–1492
Zoroaster ophiurus Fisher, 1905	1609–2227

Family Asteriidae

Evasterias troschelii (Stimpson, 1862)	0–75
Leptasterias aequalis sp. complex	intertidal to shallow subtidal
Leptasterias alaskensis sp. complex	intertidal to shallow subtidal
Leptasterias coei Verrill, 1914	18–187
Leptasterias hexactis sp. complex	intertidal to shallow subtidal
Leptasterias polaris katherinae (Grey, 1840)	0-10
Lethasterias nanimensis (Verrill, 1914)	0–224
Orthasterias koehleri (de Loriol, 1897)	0–230
Pisaster brevispinus (Stimpson, 1857)	0–128
Pisaster ochraceus (Brandt, 1835)	0–97
Pycnopodia helianthoides (Brandt, 1835)	0–120 (?435)
Stephanasterias albula (Stimpson, 1853) [ND]	33–2300
Stylasterias forreri (de Loriol, 1887)	6–532

Family Pedicellasteridae

Ampheraster marianus (Ludwig, 1905)	507–1237
Pedicellaster magister Fisher, 1923 [ND]	22–1402
Tarsaster alaskanus Fisher, 1928 [ND]	198–2100

Family Labidiasteridae

Rathbunaster californicus Fisher,1906 [SD]	99–768

Order Brisingida

Family Brisingidae

Astrocles actinodetus Fisher, 1917	2871–4200
Astrolirus panamensis (Ludwig, 1905) [SD]	48–2418
Craterobrisinga synaptoma Fisher, 1917	768–2906

Family Freyellidae

Freyella microplax (Fisher, 1917)	1576–2906
Freyellaster fecundus (Fisher, 1905)	1100–1944

SPECIES ACCOUNTS

The species accounts are arranged taxonomically by Order and Family, and the species within each Family are ordered alphabetically. Each species account is divided into the following sections.

Species Names

The current scientific name is followed by the generally accepted common name(s) and the derivation of the species name. I did not create any new common names.

Description

This section describes the animal's general appearance and the details of the external surface; subspecies descriptions follow the main one. Species are illustrated with a black-and-white photograph of a dried specimen, line drawings of important details and a colour photograph (in a separate section). The shallow-water species were photographed, where possible, in their natural habitat, deep-water species were photographed live in an aquarium or after being preserved and dried. For each species, I describe the living colour and possible variations, and in some cases, the colour in alcohol.

Size is indicated by the distance from the centre of the disc to the tip of an arm, referred to as the arm length or radius (R) of the sea star. The ratio of the arm length (R) to the radius of the disc (r) describes the animal's proportions (see figure 3). In the species accounts, this ratio is denoted by the R value, assuming r = 1. For example, an arm-to-disc ratio of 2 means that the arm is twice as long as the disc radius. For many species, the ratio is variable, so I

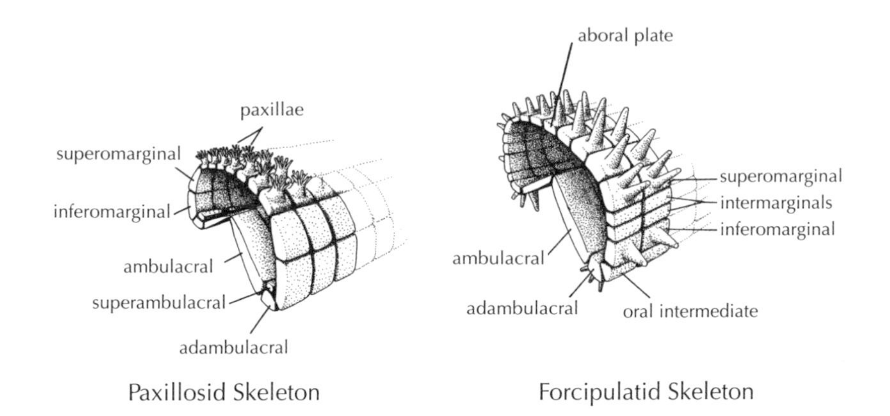

27. Calcareous plates of two skeletal types.

give a range of values. A species with an arm-to-disc ratio of 4.0 to 5.5 can have arms that are from four times as long as the disc radius to five-and-a-half times as long.

For some specimens, you may need a magnifying glass or dissecting microscope to identify taxonomically important details, such as the arrangement of spines and plates, and details of the paxillae, gills and pedicellariae. Most sea stars have a regular series of marginal plates on the side or edge of the arm (figure 27) made up of two rows, superomarginals and inferomarginals, and sometimes a row of intermarginals between them. Marginals are counted from the arm tip to half way between the two arms. (In other publications, the two rows of marginals may be designated by the prefixes supra- and infra-, superior and inferior, or upper and lower.)

Oral intermediates (also called actinal intermediates) occur between the inferomarginals and the adambulacrals (figures 27 and 28). These may form one or more rows along the arm, or be confined to a triangular region called the oral interradial area (figure 28) bounded by adjacent ambulacral furrows and the edge of the disc.

Adambulacral plates form the edge of the ambulacral furrow (figures 27, 28 and 29). Each plate bears a characteristic set of spines. Furrow spines sit on the edge of the plate next to the furrow. Spines on the oral surface of the plate may be in a longitudinal series parallel to the furrow, or a transverse series at right angles to

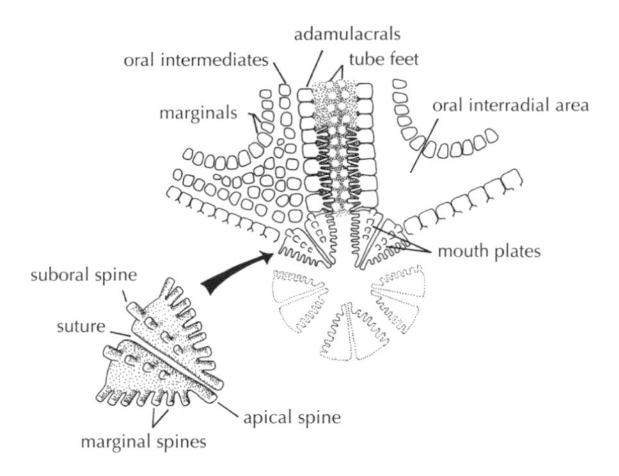

28. Calcareous plates of the oral side.

the furrow. The length of a plate is measured parallel to the ambulacral furrow or arm axis, and the width is measured at right angles to that; this is why the width of a plate may be greater than its length.

Triangular mouth plates (also mouth angle plates or mouth angle ossicles) are situated at the proximal end (nearest the mouth) of the adambulacrals (figure 28). Marginal spines line the free edge of the mouth plate. The number of marginal spines in the description refers to those on one plate of the pair from the apical spine, nearest the mouth, to the spine closest to the first adambulacral plate. Spines on the oral surface of the plate are called suborals. In most species of Forcipulatida, a few proximal adambulacrals from adjoining arms fuse together into an adoral carina (figure 110).

Some Descriptions end with a Taxonomic Note, which shows the original scientific name and author, and the author who proposed the present name, as well as other relevant comments.

Similar Species

This section flags other species in the region that are similar in appearance, and highlights the distinguishing characteristics. Simply compare the description of this species with those of the similar ones.

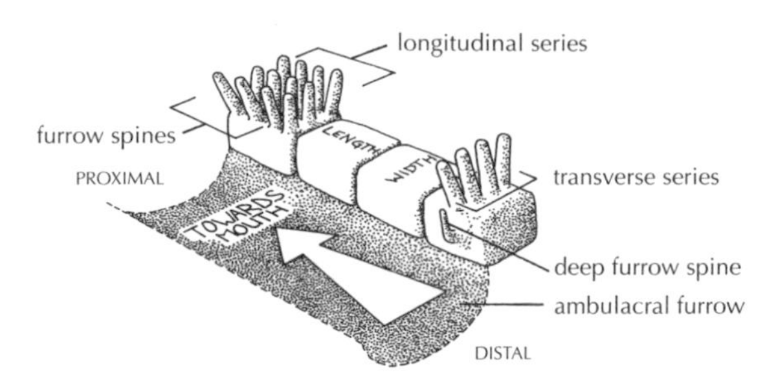

29. Adambulacral plates.

Distribution

Here I give the geographic range and habitat of the species. The range is the species' known northern and southern limits, and the shallowest and deepest depths that it is found in. Maps for each species are unnecessary because all the species are distributed in a relatively narrow band along the coast between the limits indicated. Throughout the text, "this region" refers to the area covered by this book, from Glacier Bay to Puget Sound to a depth to 200 metres. The habitat is the substrate most often associated with the species. I also include my estimate of the species' abundance in this region, based on my intertidal and subtidal observations and on its frequency in museum collections.

Biology

This section includes feeding, breeding and other biological information, such as parasites, behaviour and physiology. Most of the information here I gleaned from published literature (listed in References), personal communications and my own observations.

References

To make the text read more smoothly for a general audience, I cite references here rather than throughout the account. Publications that I based my geographic range on are listed separately. This allows easy access to the original references for researchers who find a specimen that extends the range or who want to challenge my data. The full citations are at the end of the book in alphabetical order by author.

ORDER PAXILLOSIDA

Family Luidiidae

Characteristics: Small superomarginals, hardly distinguishable from the adjacent aboral paxillae; large, well-developed inferomarginals, with long spines. Oral intermediates almost to the tip of the arm. Pointed tube feet. Branching papulae. Flat disc and arms. No anus, intestine or intestinal caeca.

In this region:
Luidia foliolata

30. *Luidia foliolata*, R = 205 mm. (See colour photograph C-2.)

Luidia foliolata **Sand Star**

foliolata: **Latin *foliolate*, leaflets, possibly for the leaflike arms.**

Description

Luidia foliolata is a long-armed, dull grey or brown sea star often partially buried in soft substrates. It has five arms up to 30 cm long, rectangular in cross-section. The arm-to-disc ratio is 3.4 to 7.1. The aboral surface is flat and the paxillae are of even height. The entire body is grey or has yellowish-white to light orange mottling; the tube feet are yellow to orange. The aboral surface consists of small, irregular paxillae down the centre of the arm, more square towards the edge; no spines or pedicellariae. Three vertical, flattened spines on the inferomarginal plates outline the arm when viewed from above; the remainder of the plate has smaller, flattened spines with fine spinelets around its perimeter (figure 31). The row above the inferomarginals, normally the superomarginals, cannot be distinguished from the adjacent aboral paxillae. A single row of small, round oral intermediate plates bear a tuft of 5 spinelets, difficult to see unless all spines are removed. A transverse series of 3 to 5 slender spines are on the oral surface of the adambulacrals (figure 31). The tube feet are large and pointed. The mouth plates are narrow but more elevated than adjacent adambulacrals. On the vertical edge nearest the mouth is a series of about 20 small spinelets; one or two series of larger suboral spinelets run the length of the plate. A group of 5 spinelets overhang the suture at the distal end of the mouth plate.

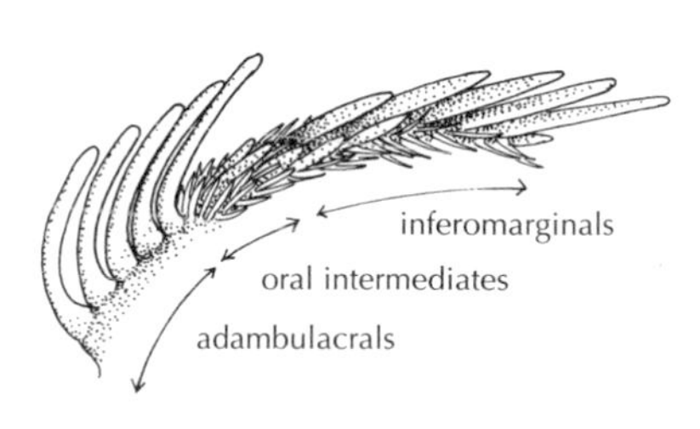

31. Spines of *L. foliolata.*

Similar Species

L. foliolata cannot be mistaken for any other species in this book.

Distribution

Cook Inlet, Alaska, to Nicaragua, 4 to 613 metres. Usually found on mud, sand or broken shells. Common on shallow soft bottoms.

Biology

Luidia foliolata feeds on bivalves, heart urchins, sea cucumbers, brittle stars, polychaete worms, crustacea and tusk shells. A common prey is the cockle *Clinocardium nuttallii*, which buries itself just under the surface of the mud. The size of prey ingested is limited by the diameter of the mouth. The brittle star *Ophiura lutkeni* exhibits a strong escape response in the presence of *L. foliolata* (W.C. Austin personal communication), and the California Sea Cucumber (*Parastichopus californicus*) crawls away while rapidly raising its anterior and posterior ends.

The breeding season of *L. foliolata* is not known, but eggs (160 to 180 micrometres) have been obtained from adults in April, August and January. Under laboratory conditions, bipinnaria larvae have been cultured for four months up to a size of 3 mm before metamorphosis. *L. foliolata* can move relatively quickly along the bottom. It loses arms readily when disturbed or handled. The scale worms *Arctonöe pulchra* and *A. vittata* are commensal on *L. foliolata*.

References

Birkeland et al. 1971, 1982, Carey 1972, Cavey and Sokoluk 1987, Cavey and Wood 1991, Ferguson 1994, Francour 1997, Hopkins and Crozier 1966, Lissner and Hart 1996, Margolin 1976, Mauzey et al. 1968, Sloan and Robinson 1983, Strathmann 1971.
Range: Alton 1966, Clark 1989, Lambert 1999.

Family Astropectinidae

Characteristics: Well developed marginal plates, the two series matching, covered with granules or spines. Small, crowded paxillae on the aboral surface. Usually a deep vertical channel (fasciole) between the marginals, serving as a ciliated food groove. Pointed tube feet with double ampullae. The anus can be well developed, small or absent.

In this region:
Leptychaster anomalus
Leptychaster arcticus
Leptychaster pacificus

Deeper than 200 metres:
Dipsacaster anoplus
Dipsacaster borealis
Leptychaster inermis
Psilaster pectinatus
Thrissacanthias pencillatus

Astropecten verrilli has been recorded in Washington, but a single record from British Columbia is suspect.

Leptychaster anomalus

anomalus: **Latin, unusual.**

Description

Leptychaster anomalus is a small sea star with five arms up to 3.4 cm long, obvious marginals but no obvious spines. It is yellowish-white when dried. The ratio of arm to disc ranges from 1.3 to 2.0. Paxillae on the aboral surface are packed close together, smaller at the centre of disc and tip of the arm, each has 40 to 45 short, round-tipped spinelets. *L. anomalus* has 10 to 18 marginal plates. Its superomarginals, wider than their length, are covered with granuliform spinelets of even size; the inferomarginals have longer spinelets. The oral intermediates form V-shaped rows. The series adjacent to the adambulacrals extends three-quarters of arm length. The adambulacrals are square with a furrow series of 4 spinelets. The oral surface has three longitudinal series of smaller spinelets, 3 to 5 in each series. The tube feet are pointed. The mouth plates have 6 or 7 marginal spines and a similar number of suboral spines along the median suture and remainder of the plate. In overall appearance, the mouth plates bristle with spinelets (figure 32).

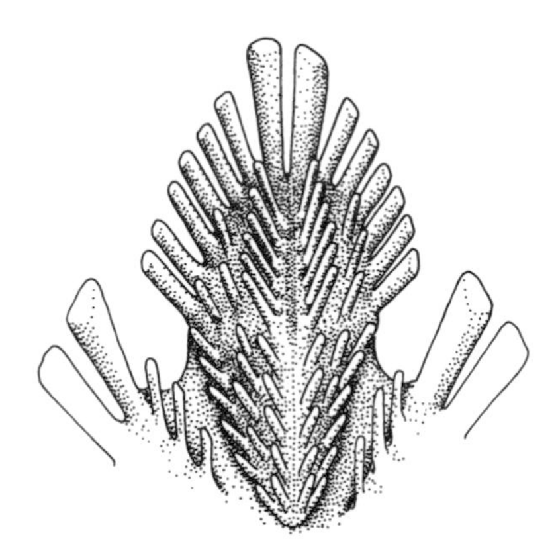

32. Mouth plates of *L. anomalus.*

Similar Species

Differs from the other two species of *Leptychaster* mainly in the arm-to-disc ratio, *L. anomalus* having shorter arms.

Distribution

The Bering Sea to Vancouver Island and to the Sea of Japan at depths of 59 to 1258 metres. Found on fine grey or black sand, green mud or pebbles. Uncommon in this region.

References

D'Yakonov 1950. **Range:** Lambert 1978b.

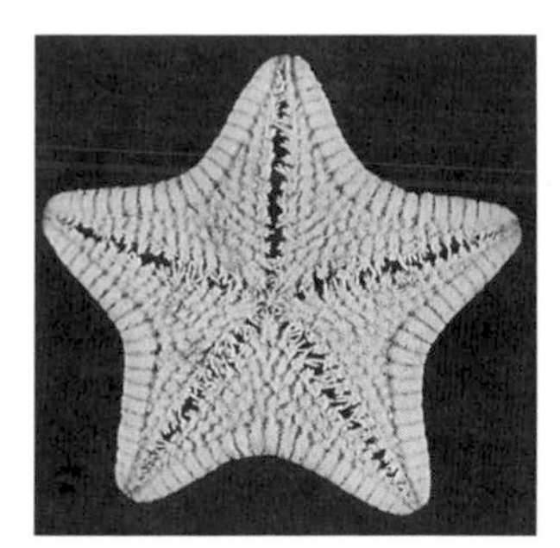

33. *Leptychaster anomalus,* R = 18 mm.

Leptychaster arcticus

arcticus: **from the type locality in the Arctic.**

Description

Leptychaster arcticus is a small pink or orange sea star with five arms up to 5 cm long, usually dredged from soft substrates. The arm-to-disc ratio is 2.0 to 3.2. Paxillae on the aboral surface are packed tightly, with compact crowns of 40 to 50 spinelets. Papulae are absent from the centre of the disc and along the centres of the arms. *L. arcticus* has 23 to 35 marginals. The superomarginals are only slightly larger than the adjacent aboral paxillae. The inferomarginals are much shorter than their width and obliquely oriented to the arm axis (figure 35). When the spinelets are removed, the groove between the plates is twice as wide as each plate. One series of oral intermediates extends along one-half the length of the arm; the second series is one-quarter of an arm's length; and a third is one-eighth or less. The adambulacrals have a curved series of 3 or 4 furrow spinelets. On the oral surface, spinelets decrease in size away from the furrow. The tube feet are pointed. The mouth plates have 10 marginal spinelets with an irregular series of smaller spinelets along the suture.

Taxonomic Note: Originally described as *Astropecten arcticus* Sars, 1851; revised to *Leptychaster arcticus* by Sladen (1889).

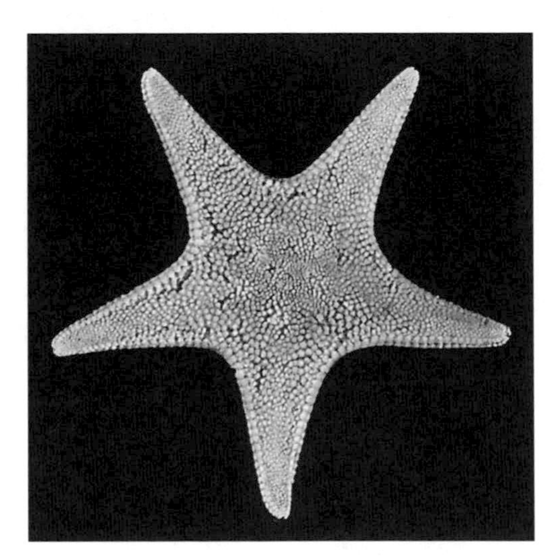

34. *Leptychaster arcticus*, R = 23 mm.

Similar Species

Leptychaster arcticus has longer arms than *L. anomalus* and smaller superomarginal plates than *L. pacificus.*

35. Marginal plates of *L. arcticus* (superomarginals above and inferomarginals below).

Distribution

Circumpolar. In the Atlantic, *Leptychaster arcticus* ranges south to 38°N on the North American side and south of Ireland on the coast of Europe. In the Pacific, it ranges from the Bering Sea to Yezo, Japan, and to northern Oregon, in depths of 40 to 1261 metres. It is found on mud, sand and pebbles. Uncommon in this region.

References

D'Yakonov 1950, Grainger 1966. **Range:** Alton 1966, Fisher 1911, Lambert 1978a, Mortensen 1927.

Leptychaster pacificus

pacificus: **refers to the species' normal range in the Pacific Ocean.**

Description

Leptychaster pacificus is a small, light pink sea star with marginals wider than their length in the interradial region and less so towards the arm tip. It becomes grey-white when preserved. It has five arms up to 4.5 cm long, and the arm-to-disc ratio ranges from 2.6 to 3.0. The aboral surface is made up of closely packed paxillae decreasing in size towards the centre of the disc. The larger paxillae have about 25 peripheral and 30 central spinelets. About 30 superomarginals are wider than their length in the interradial area and gradually become squarish distally; the inferomarginals correspond to the superomarginals but are two to three times wider and obliquely oriented to the furrow. The oral interradial area is similar to *L. arcticus* or slightly smaller. The adambulacrals are squarish with 4 or 5 slender, blunt furrow spinelets and two or three longitudinal series of about 4 spinelets on the oral surface. The tube feet are pointed. The mouth plates have about 15 slender marginal spinelets.

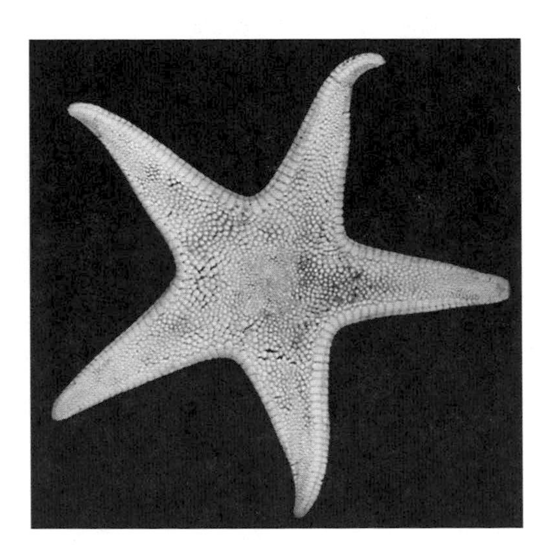

36. *Leptychaster pacificus*, R = 43 mm. (See colour photograph C-3.)

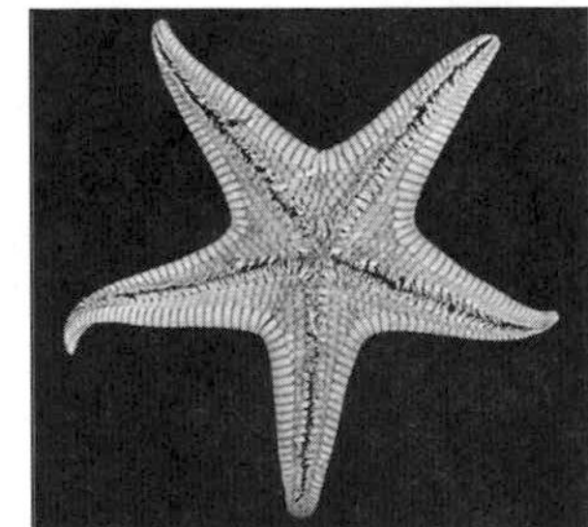

Similar Species

Leptychaster pacificus has more obvious marginals than *L. arcticus*. Its arm-to-disc ratio is greater than that of the short-armed *L. anomalus*.

Distribution

From the southern Bering Sea to Cape Flattery, Washington, in depths of 10 to 435 metres. It is found on soft mud, sand or pebbles. Common in this region.

References

McDaniel et al. 1978. **Range:** Lambert 1978b.

Family Porcellanasteridae

No species in this region.

Deeper than than 200 metres:
Eremicaster crassus
Eremicaster pacificus

Family Ctenodiscidae

Characteristics: Ciliated grooves (cribriform organs) between the marginals, which continue across the oral surface as grooves between transverse rows of plates edged with spinelets. The aboral surface has paxillae, and its marginals are moderately solid. Pointed tube feet. No intestine.

In this region:
Ctenodiscus crispatus.

37. *Ctenodiscus crispatus,* R = 25 mm. (See colour photograph C-4.)

Ctenodiscus crispatus **Mud Star**

crispatus: **Latin *crispus*, curly.**

Description

Ctenodiscus crispatus is a small sea star, with a soft, slightly inflated aboral surface bearing a characteristic elevated cone in the centre (epiproctal cone). Five short triangular arms (exceptionally four or six) extend up to 5.4 cm long. The aboral side is grey to yellow-white, and the oral side is lighter with tinges of pale orange. The ratio of arm to disc is from 1.7 to 2.7. The paxillae of the aboral surface are low, with few to many short skin-covered spinelets. Each arm has a definite vertical side formed by 11 to 20 marginals. There is a single conical spine at the upper ends of the superomarginals and inferomarginals (figure 38). Deep grooves between the marginals and across the oral intermediates are bordered on both sides by flat spinelets that extend over the furrow. The adambulacrals have an oblique series of 3 to 5 sharp furrow spines. The tube feet are large and pointed. The mouth plates are prominent, with 7 marginal spines; the apical spine is the largest (figure 38).

Taxonomic Note: Originally described as *Asterias crispata* Retzius; revised to *Ctenodiscus crispatus* by Düben and Koren (1846).

Similar Species

No other species has an elevated cone in the centre of the disc.

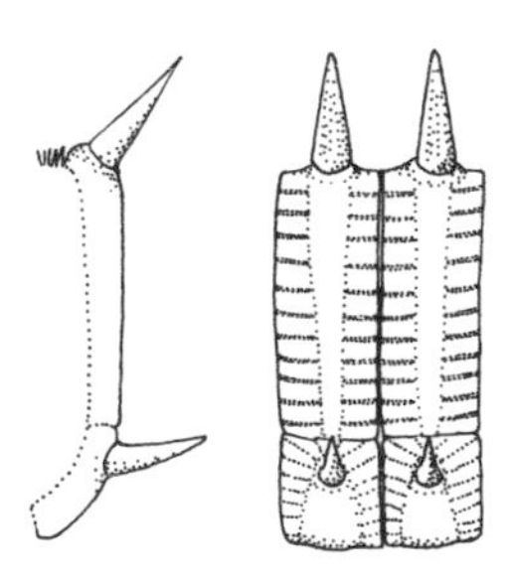

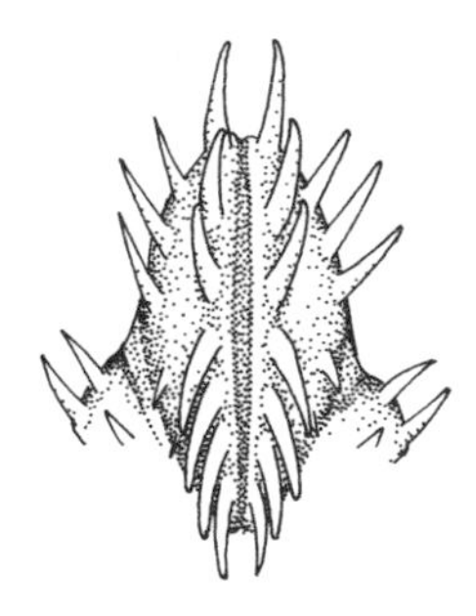

38. Plates of *C. crispatus* (from left to right): side and front views of superomarginals (above) and inferomarginals (below); mouth plates.

Distribution

Circumpolar to New England on the Atlantic coast, and to Panama and Japan in the Pacific, at depths of 10 to 1890 metres. Found on soft mud and rock, and in the Arctic, on rock or sand. Common in this region.

Biology

Ctenodiscus crispatus is a non-selective deposit feeder. Tube feet around the mouth shovel sediment into the stomach, which digests the organic material, much of it bacteria on the surface of sand grains. Fragments of bivalves, small crustaceans and worm tubes have been found in the stomachs of collected specimens. *C. crispatus* regularly extrudes indigestible material. It establishes a temporary burrow in soft sediment and uses its extensible epiproctal cone to maintain a connection with the water above. Currents created by ciliated grooves called cribriform organs draw sea water into the burrow for respiration. Any incidental food particles are trapped by a cleansing mucus and ingested. Cells lining these grooves have complex surfaces (microvilli) that can absorb amino acids directly from sea water. This type of nutrition may be important for surface cells that are not supplied by an internal blood system.

In the Gulf of Maine, there is no seasonal variation in gonad size, suggesting continuous reproduction. The full range of eggs can be found in the gonad at any one time, but large mature eggs are most numerous in late autumn and mid winter in a Norwegian fiord. As with some other invertebrates, spawning seems to be related to phytoplankton production. The large, yolky egg implies direct development, but this is not known for certain. The Mud Star attains full size in three years. The species is extremely variable genetically.

References

Alton 1966, Bell and Sargent 1985, Carey 1972, Falk-Petersen 1982, Falk-Peterson and Sargent 1982, Grainger 1966, Komatsu 1986, McDaniel et al. 1978, Shick 1976, Shick et al. 1981a and 1981b, Turner and Dearborn 1972, Walker 1979. **Range:** D'Yakonov 1950, Fisher 1911.

ORDER NOTOMYOTIDA

Family Benthopectinidae

Characteristics: Primarily a deep water family. Small disc; long, narrow arms that taper towards the tips. The main characteristic is a pair of longitudinal muscles on the aboral side of each arm from the level of the third to ninth superomarginal to the arm tip. Papulae confined to papularia on the arms or disc. Supero- and inferomarginals alternate and have long spines. The tube feet have suckers. The pedicellariae are pectinate.

In this region:
Cheiraster dawsoni

Deeper than 200 metres:
Benthopecten claviger claviger
Benthopecten mutabilis.
Nearchaster aciculosus
Nearchaster variabilis
Pectinaster agassizi evoplus

Cheiraster dawsoni

dawsoni: **after Dr G.M. Dawson, Geological Survey of Canada.**

Description

Cheiraster dawsoni is a spiny vermilion sea star with five arms up to 17 cm long. The ratio of arm to disc is from 5.0 to 5.5. The larger aboral plates bear a large central spine surrounded at the base by a circle of 2 to 8 unequal spinelets; the smaller plates bear 1 to 12 spinelets (figure 39, top right). Papulae are absent from the centre of the disc, the aboral interradial area and the outer two-thirds of the arm; there are 3 to 8 pectinate pedicellariae on proximal half of each arm (figure 39, top left). The superomarginals and inferomarginals are staggered (not opposite each other). The superomarginals bear 2 stout tapering spines, and around each spine is a circle of 12 to 15 slender spinelets. The inferomarginals have 3 or 4 rigid tapering spines surrounded by auxiliary spinelets that are one-third the length of the spine. The oral interradial area has a few plates with small, well-spaced spinelets, 1 or 2 tapering spines and 1 to 3 pectinate pedicellariae. The adambulacrals are wider than their length; the curved furrow margin bears 4 to 7 spinelets, of which the central 3 or 4 are longer than the width of the plate; the oral surface has a transverse series of 2 or 3 long, slender spines, with a few spinelets around the edge (figure 39, bottom). Mouth plates have 6 to 8 marginal spines, the apical spine being the largest, and 7 to 10 suborals.

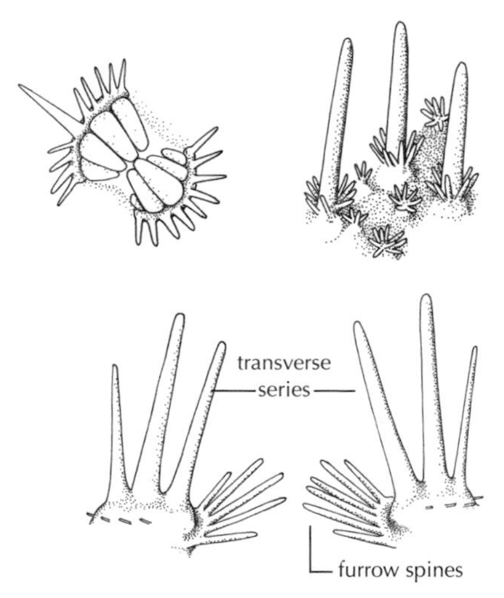

39. Identifying features of *C. dawsoni*: pectinate pedicellaria (top left), aboral spines (top right) and adambulacral spines (bottom).

Taxonomic Note: Originally described as *Archaster dawsoni* Verrill; revised to *Cheiraster (Luidiaster) dawsoni* by A.M. Clark (1981).

40. *Cheiraster dawsoni*, R = 130 mm. (See colour photograph C-5.)

Similar Species

Above 200 metres depth, this species is very distinctive in shape and surface detail. In depths greater than 200 metres it might be confused with *Nearchaster* or *Benthopecten* species.

Distribution

The Bering Sea to northern Oregon in depths of 73 to 384 metres. Found on mud, sand, pebbles and rock. Common.

References

D'Yakonov 1950. **Range:** Alton 1966, Lambert 1978a.

ORDER VALVATIDA

Family Asterinidae

Characteristics: Flat body with short arms. The edge of body is sharp; the marginals are minute. The aboral surface is covered with overlapping plates bearing granules or short spinelets. The oral interradial area is also covered with overlapping plates bearing spinelets.

In this region:
Asterina miniata

41. *Asterina miniata*, R = 70 mm. (See colour photograph C-1.)

Asterina miniata **Bat Star**

miniata: **Latin *miniatus*, bright red.**

Description

Asterina miniata is a broad-armed sea star found in large numbers in shallow, sheltered waters on the west coast of Vancouver Island, the north coast of British Columbia, and the Queen Charlotte Islands. Its colour is extremely variable, and can be mottled or uniform red, yellow, brown, green and blue. The five arms (occasionally six) are up to 9 cm long, and it has an arm-to-disc ratio of 1.6 to 2. The aboral plates along the radii of each arm are low and crescent-shaped with their concave side towards the centre (figure 42). Smaller irregular plates are topped with a circle of granules. A row of tiny inferomarginals form the sharp edge between the aboral and oral sides. Immediately above them is a row of even smaller, round superomarginals. The oral interradial area is relatively flat from the mouth to the edge of the disc. Oral intermediates form a V-shaped pattern with the apex towards the mouth. Each plate has a webbed comb of 3 to 5 flat spines pointing away from the mouth (figure 43) that get smaller distally. The adambulacral plates have 2 to 4 furrow spines and an oblique row of 2 or 3 similar-sized spines on the oral surface. The mouth plates have 5 marginal spines and 2 or 3 heavier scoop-shaped suborals.

Taxonomic Note: Originally described as *Asterias miniata* Brandt; revised to *Asterina miniata* by Sladen (1889); also known as *Patiria miniata*.

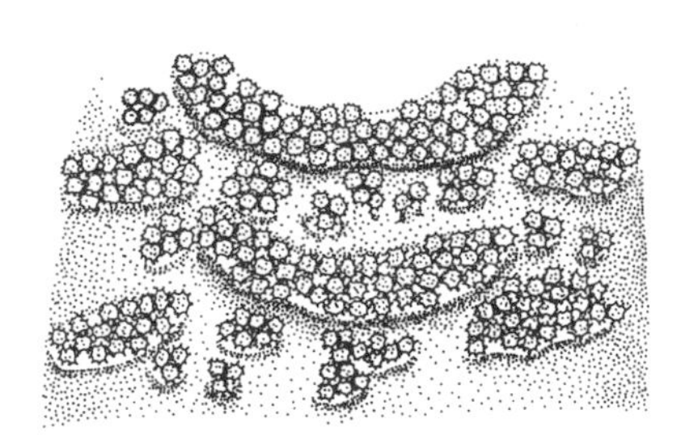

42. Aboral plates of *A. miniata*.

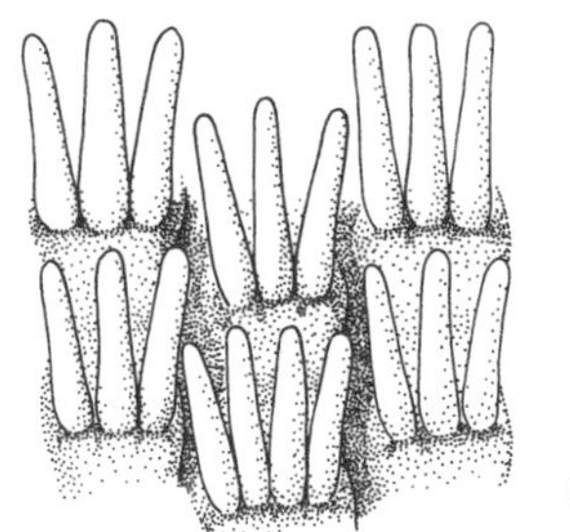

43. Oral intermediate plates.

Similar Species

A. miniata has a similar shape to *Dermasterias imbricata*, but it feels like sandpaper rather than smooth and slippery.

Distribution

Sitka, Alaska, to the Gulf of California; intertidal to 302 metres. *Asterina miniata* is abundant intertidally on rocks, broken shells, gravel and sand on the exposed coast of British Columbia, but not in areas under the direct influence of the surf. It is found on the protected side of Vancouver Island only as far as the Gulf Islands in the south and the Port Hardy region in the north.

Biology

The diet of *Asterina miniata* includes dead animals, seaweed, sponges, sea urchins and squid eggs. In southern California it feeds on the bryozoan *Tubulipora,* one of the first species to colonize a surface. Removing that bryozoan allows successional species to colonize sooner. *A. miniata* has an extremely large cardiac stomach with elaborate retractor muscles, but it is not able to open bivalves. Most of the digestive enzymes are produced by the pyloric caeca rather than the stomach as in other species. There is some evidence that it feeds on suspended particles with the aid of mucus. Bat Stars reach an arm length of 4 cm in about three years.

Populations of *A. miniata* may breed at any time of year, but individuals are not synchronized with each other. Each has an annual reproductive cycle requiring two months to rebuild the gonad after spawning. In California, the majority spawn in late spring or early summer and there may be another spawning in August. In Barkley Sound, recruitment originated mainly from the local population rather than from larvae migrating in from other regions. The final pattern of settlement was controlled by events just before settling or just after. For example, juvenile *A. miniata* were attacked by *Pycnopodia helianthoides*. *Ophiodromus* (formerly *Podarke*) *pugettensis,* a commensal scale worm, is often found on *A. miniata*. The structure of the eye spot is described by Eakin and Brandenburger (1979).

References
Anderson 1959 and 1960, Barber 1979, Brandenburger and Eakin 1980, Bruno et al. 1992, Cameron and Holland 1983, Chaffee and Spies 1982, Chiba et al. 1990, Clark 1983, d'Auria et al. 1988, 1990a and 1990b, Davis 1985, Davis et al. 1981, Day and Osman 1981, Eakin and Brandenburger 1979, Farmanfarmaian et al. 1958, Heath 1917, Holland 1980, Hopkins and Crozier 1966, Horowitz et al. 1994, Huvard and Holland 1986, Khotimchenko et al. 1985 and 1987, Kishimoto et al. 1982, Lopo and Vacquier 1980, Meijer and Guerrier 1981, Meijer and Wallace 1980, Rosenberg 1979, Rosenberg and Lee 1981, Rumrill 1989, Schroeter et al. 1983, Spies and Davis 1982, Steffen and Linck 1988, Strathmann 1971, Stricker et al. 1994a, Tegner and Dayton 1987, Thomas 1981, Van Veldhuizen and Oakes 1981, Webster 1975, Wobber 1975.
Range: Fisher 1911.

Family Poraniidae

Characteristics: Short armed and star shaped or almost pentagonal in form; the upper side is arched, the lower side flat. The aboral plates form a compact or open reticulum, but are obscured by a thick, opaque skin. Some have carinals linked to the marginals by an open reticulum with large nodal plates (as in *Poraniopsis*); only *Poraniopsis* has spines at each node. The marginals vary from prominent to small or absent; they are covered with skin and usually have no prominent spines. The oral intermediates run parallel to the inferomarginals. The adambulacrals have a few sheathed spines. Spaced or clustered papulae; no pedicellariae.

In this region:
Poraniopsis inflatus inflatus

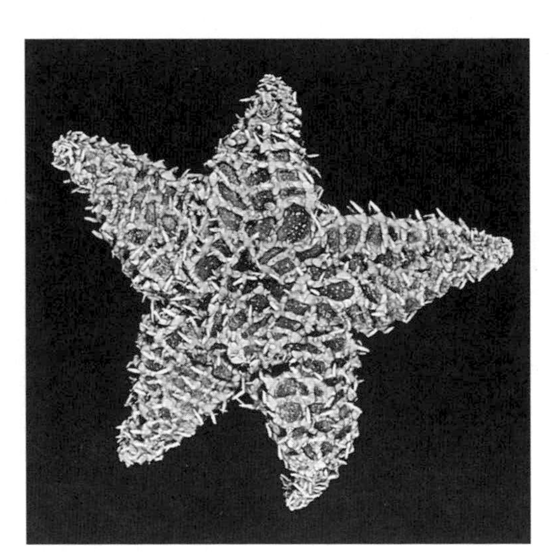

44. *Poraniopsis inflatus inflatus,* R = 73 mm. (See colour photograph C-6.)

Poraniopsis inflatus inflatus

inflatus: **Latin, puffed up, swollen.**

Description

Poraniopsis inflatus inflatus is cream to bright orange with large white aboral spines. The five fat arms are up to 10 cm long and the arm-to-disc ratio ranges from 2.2 to 3.0. The aboral plates form an open mesh of large squarish areas occupied by numerous papulae. Prominent pointed spines occur at the junctions of the plates and a few smaller spines occur on secondary plates (figure 45). Marginals are not readily distinguishable, because both series resemble the aboral plates and spines. Usually a row of smaller intermarginal spines are found on transverse plates between the two marginal series. Oral intermediates bear a few smaller, scattered spines between the inferomarginals and adambulacrals. The adambulacrals have one flat furrow spine and a similar flat or scoop-shaped spine on the oral surface. The mouth plates bear four marginal spines, the apical spine being largest and similar in size and shape to the adambulacral spines.

Taxonomic Note: Originally described as *Alexandraster inflatus* Fisher; revised to *Poraniopsis inflata* by Fisher (1910) and to *Poraniopsis inflatus inflatus* by Clark (1993).

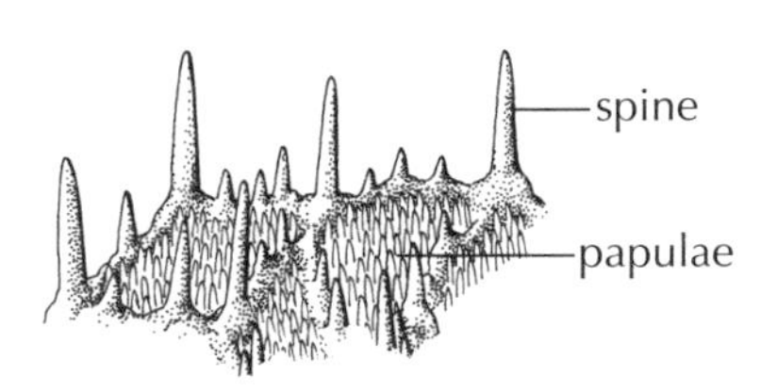

45. Aboral surface of *P.i. inflatus.*

Similar Species

Difficult to confuse with any other species of sea star.

Distribution

The Gulf of Alaska, and from the Queen Charlotte Islands to San Diego, California; also to northern Honshu and Moneron Island, USSR. This sea star lives at depths of 11 to 366 metres on a variety of substrates, from mud to rocks. Uncommon in this region; only eight specimens are recorded in collections, but Neil McDaniel saw three during one dive in Barkley Sound.

Biology

Not much is known about the feeding habits of *Poraniopsis inflatus inflatus*. Captive animals seem to prefer sponges. A spacious cardiac stomach with large retractor muscles suggests that the stomach can be extruded to digest large prey. In Tasu Sound, Queen Charlotte Islands, I collected two specimens among expanses of the large solitary cup corals *Paracyathus stearnsi*, which might be a likely prey. *Dendrogaster punctata*, a primitive barnacle, parasitizes *P.i. inflatus*.

References

Alton 1966, Anderson and Shimek 1993, Clark 1984, D'Yakonov 1950, Grygier 1982, Hopkins and Crozier 1966, Lissner and Hart 1996, Oguro 1989. **Range:** Lambert 1978b and 1999.

Family Goniasteridae

Characteristics: Flat body. Prominent marginals; the supero- and inferomarginals are similar. The aboral plates are flat or paxilliform, bearing granules, low stumps or spines. Only the radial areas have papulae. Pedicellariae are generally present. The tube feet have suckers.

In this region:

Ceramaster arcticus
Ceramaster patagonicus
Gephyreaster swifti
Hippasteria spinosa
Mediaster aequalis
Pseudarchaster alascensis

Deeper than 200 metres:

Ceramaster clarki
Ceramaster japonicus
Cryptopeltaster lepidonotus
Hippasteria californica
Pseudarchaster dissonus
Pseudarchaster parelii

Ceramaster arcticus **Arctic Cookie Star**

arcticus: **first described from the Arctic (Bering Sea).**

Description

Ceramaster arcticus is small and pentagonal, stiff and firm to the touch, and up to 5.5 cm in greatest radius. It is pale orange with red patches. The ratio of arm to disc is 1.6. The aboral surface has tabulate plates bearing 4 to 12 marginal granules surrounding 1 to 3 smaller central granules (figure 47, right). *C. arcticus* has 18 to 24 massive superomarginals, each usually with a bare spot surrounded by flat granules; the inferomarginals each have a smaller bare spot; papulae cover the aboral surface. The adambulacrals have 2 or 3 short, thick furrow spinelets with 3 to 6 granules on oral surface of the plate (figure 47, left), which often has a bivalved pedicellaria.

Taxonomic Note: Originally described as *Tosia arctica* Verrill; revised to *Ceramaster arcticus* by (Fisher 1911).

Similar Species

Ceramaster arcticus differs from *C. patagonicus* in its smaller average size, fewer granules on the aboral tabulate plates and fewer adambulacral spines.

46. *Ceramaster arcticus*, R = 26 mm. (See colour photograph C-8.)

Distribution

Bering Sea to the Juan de Fuca Strait; intertidal to 186 metres. Found on mud in the deep part of its range to rock in shallow waters. Rare in this region.

References

Bavendam 1985, D'Yakonov 1950. **Range:** Lambert 1978b.

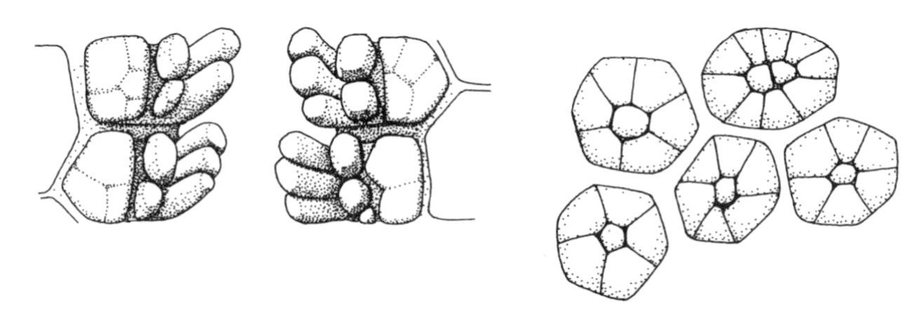

47. Adambulacral spines (left) of *C. arcticus* and the granules on its aboral tabulate plates (right).

Ceramaster patagonicus Cookie Star

patagonicus: **first described from the Straits of Magellan, near Patagonia.**

Description

Ceramaster patagonicus is pentagonal in shape and up to 8.2 cm in greatest radius, often with the aboral surface swollen and soft to the touch in living specimens. It is creamy orange to red-orange on the aboral side and pale yellow orally. The ratio of arm to disc ranges from 1.3 to 1.7. *C. patagonicus* has regular hexagonal aboral plates on the radii of the arms, and square or rhomboid plates between the radii. Each aboral tabulate plate is covered with 12 to 15 marginal granules and 4 to 12 central granules (figure 49); a few plates bear spatulate pedicellariae. The marginals are massive and the granules crowded together. The oral intermediates are four sided, with granules and pedicellariae similar to those on the aboral plates. The adambulacrals have 3 to 5 robust spinelets on the edge of the furrow; distal to these on the oral surface is a longitudinal row of 2 or 3 short, stubby spinelets and then 5 to 8 irregular granules (figure 50). The mouth plates have 8 or 9 blunt, prismatic marginal spines (figure 51).

Taxonomic Note: Originally described as *Pentagonaster patagonicus* Sladen; revised to *Ceramaster patagonicus* by Fisher (1911). Clark and Downey (1992) consider *C. patagonicus* to be a subspecies of

48. *Ceramaster patagonicus*, R = 73 mm. (See colour photograph C-9.)

Ceramaster grenadensis from the Atlantic: *C. grenadensis patagonicus*. Yet Clark (1993) suggests that the subspecies *C. patagonicus fisheri* Bernasconi, 1963, may be the form on our coast. Until this question is resolved I opted to leave it as *C. patagonicus* and not complicate the story with possible subspecies.

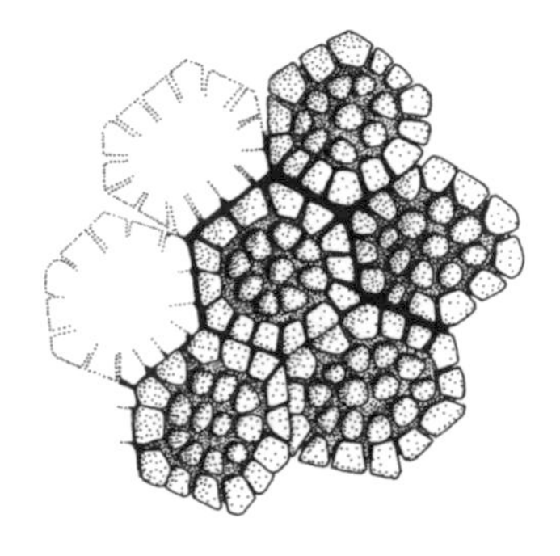

49. Aboral view of tabulate plates.

Similar Species

Ceramaster patagonicus is larger than *C. arcticus*, and it has more granules on the aboral tabulate plates and more furrow spines.

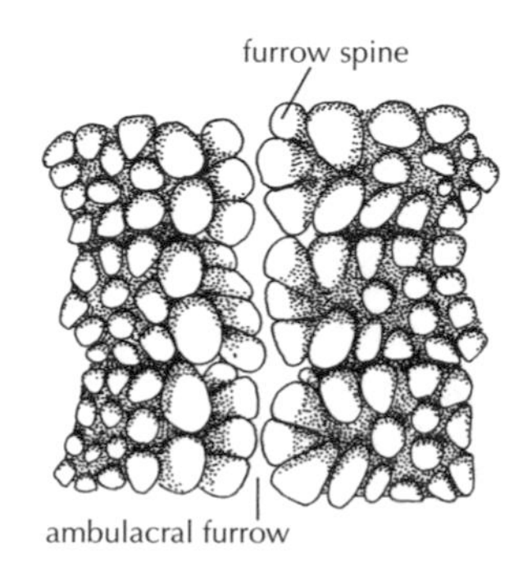

50. Adambulacral spines.

Distribution

Ceramaster patagonicus (including subspecies *patagonicus* and *fisheri*) ranges from the Bering Sea to Cape Horn, South America, in depths of 10 to 245 metres. It is found on rocks or mud. Within diving depth, it is common in some British Columbia inlets, but uncommon in the Strait of Georgia and exposed locations.

References

Clark 1993, Clark and Downey 1992, D'Yakonov 1950, Lambert 1978b. **Range:** Clark 1993, Clark and Downey 1992, Lambert 1981a and 1999.

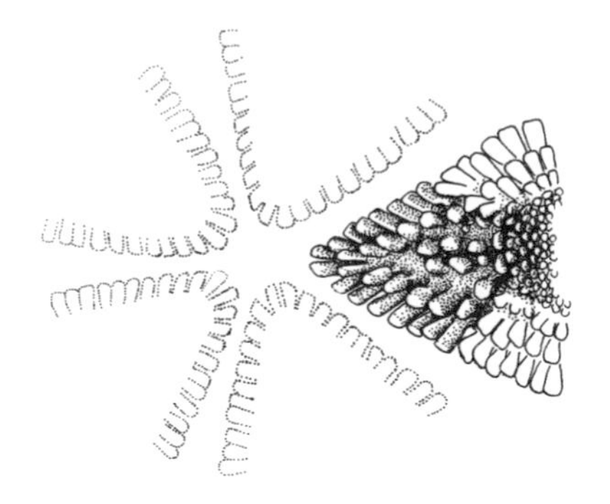

51. Mouth plates of *C. patagonicus*.

Gephyreaster swifti **Gunpowder Star**

swifti: **possibly after someone named Swift.**

Description

Gephyreaster swifti is star shaped with small, tightly packed aboral paxillae and obvious granule-covered marginals. The aboral side is pink-orange and the oral side is paler. It has five arms up to 21 cm long, and an arm-to-disc ratio of 2.1 to 2.6. Rich Mattison of Juneau, Alaska, brought me a live specimen of *G. swifti* that was 41.5 cm in diameter. The aboral paxillae are closely packed and crowned with numerous granuliform spinelets (figure 52). Between the paxillae, and almost hidden by them, are 1 to 3 papulae. The marginals are conspicuous and rounded, with a definite groove between infero- and superomarginals (figure 53). The oral interradial area has oblong plates covered with coarse spinelets that lie in rows between the inferomarginals and corresponding adambulacrals. The adambulacrals have 2 or 3 furrow spines that almost cover the ambulacral furrow; and the oral surface has two or three transverse

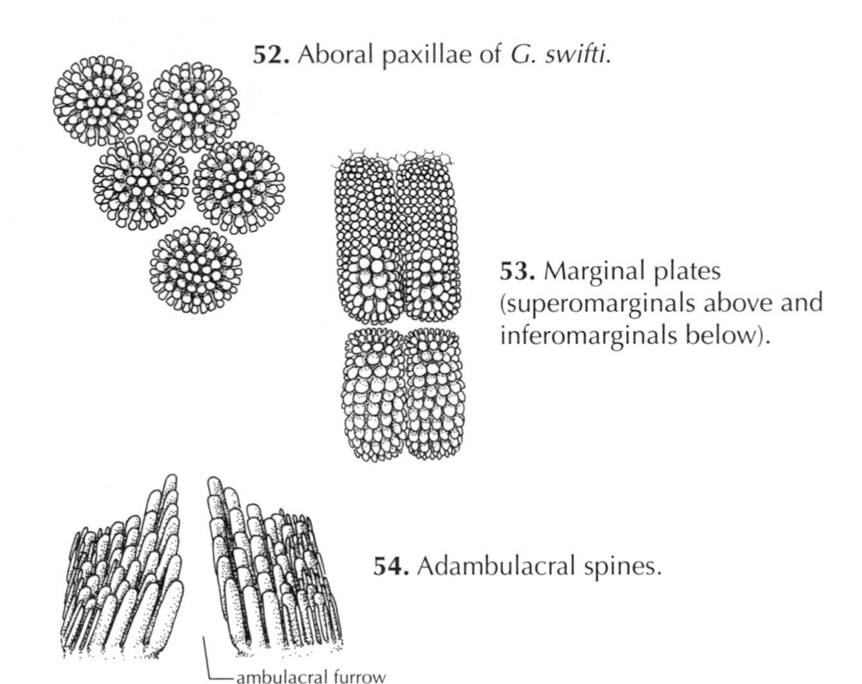

52. Aboral paxillae of *G. swifti.*

53. Marginal plates (superomarginals above and inferomarginals below).

54. Adambulacral spines.

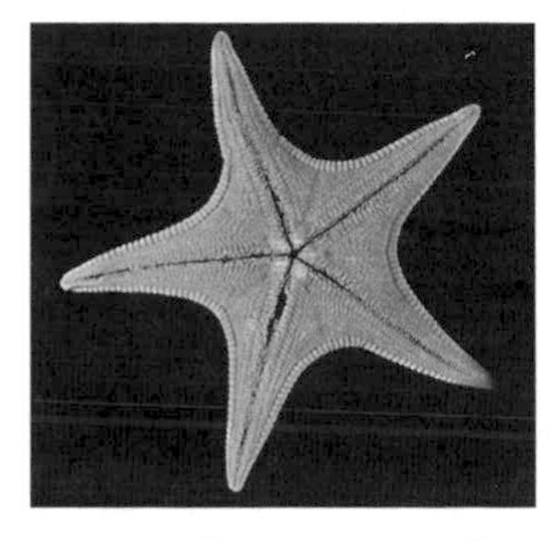

55. *Gephyreaster swifti*, R = 115 mm. (See colour photograph C-7.)

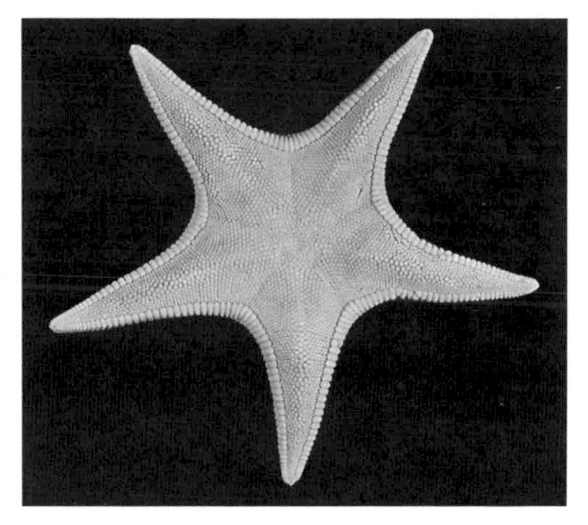

rows of 5 or 6 spines (figure 54). The mouth plates are prominent, with numerous small blunt spines on the oral surface and numerous marginal spines in a series progressing to a single large apical spine.

Taxonomic Note: Originally described as *Mimaster swifti* Fisher; revised to *Gephyreaster swifti* by Fisher (1911).

Similar Species

Gephyreaster swifti is never vermilion like *Mediaster aequalis*, which is similar in shape but has smoother marginals that encroach much more on the aboral surface. *Pseudarchaster alascensis* is flatter, smoother and its arms taper more sharply.

Distribution

Bering Sea and Aleutian Islands to Washington, in depths of 11 to 344 metres on rock or sand. Uncommon in southern British Columbia; more common in northern waters.

Biology

Feeds on sea anemones, especially *Metridium senile* and *Stomphia coccinea*.

References

D'Yakonov 1950, Lambert 1998, Mauzey et al. 1968.
Range: Fisher 1911.

Hippasteria spinosa **Spiny Red Sea Star**

spinosus: **Latin *spina*, thorn.**

Description

Hippasteria spinosa is spiny, vermilion and star-shaped with five arms up to 17 cm long. The oral surface is usually a lighter orange to white. The ratio of arm to disc ranges from 1.7 to 2.6. The aboral surface has large plates, each bearing a single stout tapering spine; secondary plates are interspersed, some with a conspicuous bivalve pedicellaria, the jaws of which are slightly wider than their height. These pedicellariae are variable and may resemble the *H. californica* type, which are taller and narrower. All plates are surrounded by granules, giving them a star-shaped appearance (figure 57). *H. spinosa* has 16 or 17 superomarginals, each surrounded by a row of smooth granules and bearing two stout spines, reducing to one at arm tip (figure 58); the inferomarginals are similar. The oral intermediates are oval and surrounded by pointed granules bearing stubby spinelets or low bivalve pedicellariae. The adambulacrals have two or three furrow spines and a single stout spine on the oral surface of the plate, surrounded by granules or spinelets (figure 59). The mouth plates have four or five marginal spines.

56. *Hippasteria spinosa*, R = 90 mm. (See colour photograph C-10.)

Similar Species

Hippasteria spinosa should not be confused with any other species in this book. Its closest relative is *H. californica* from deeper water.

57. Aboral spines of *H. spinosa*, with a bivalve pedicellaria in the centre.

Distribution

Kodiak Island, Alaska, to southern California on the North American coast and to the Sea of Okhotsk in the western Pacific. Found in depths of 10 to 512 metres on mud, sand, shell or rock. Common below 100 metres; also common at diving depth on the west coast of Vancouver Island and other exposed parts of the coast.

Biology

Hippasteria spinosa feeds primarily on the orange sea pen *Ptilosarcus*, which it consumes by everting its stomach over the apical end and progressing down the length. This sea star also eats the white-plumed anemone *Metridium*, the zoantharian *Epizoanthus scotinus*, the colonial sea squirt *Metandrocarpa*, the polychaete worm *Nereis*, and the eggs of the nudibranch *Armina*. It is also known to evoke the swimming response of the anemone *Stomphia*.

H. spinosa probably breeds from May to July, producing pelagic lecithotrophic larvae.

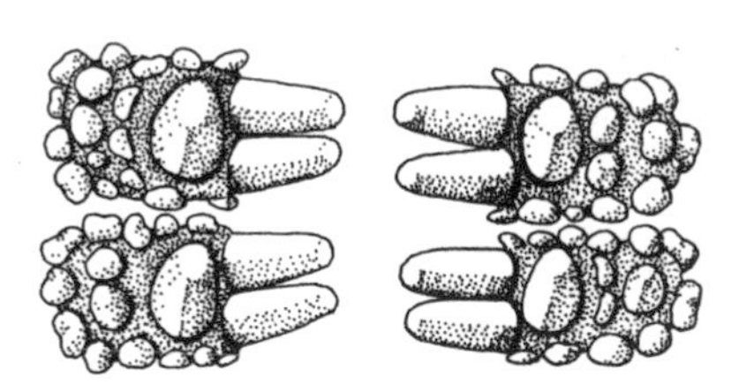

58. Adambulacral spines.

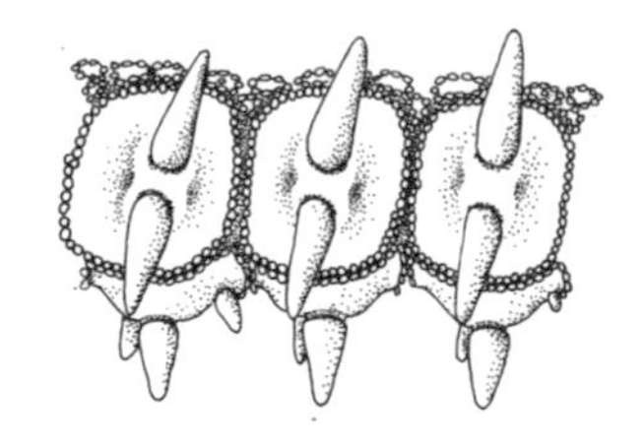

59. Marginal spines (superomarginals above and inferomarginals below).

References

Alton 1966, Carey 1972, Grygier 1982, Mauzey et al. 1968, Strathmann 1987, Ward 1965b. **Range:** D'Yakonov 1950, Fisher 1911.

60. *Mediaster aequalis,* R = 85 mm. (See colour photograph C-11.)

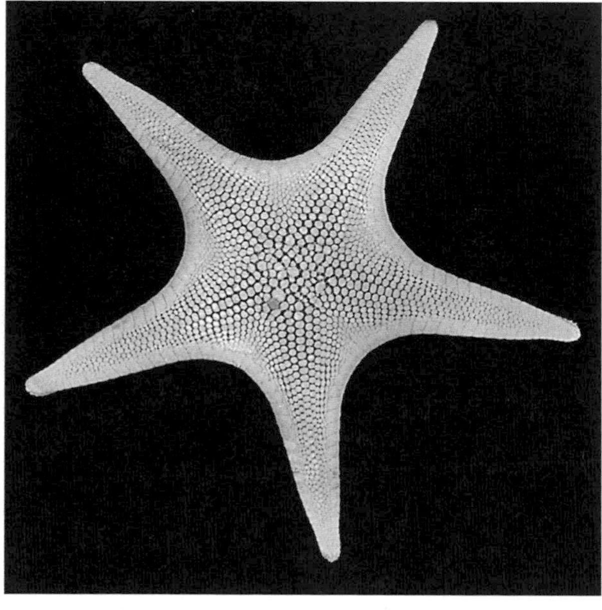

Mediaster aequalis **Vermilion Star**

aequalis: **Latin, same, uniform, referring to the regular shape of this species.**

Description

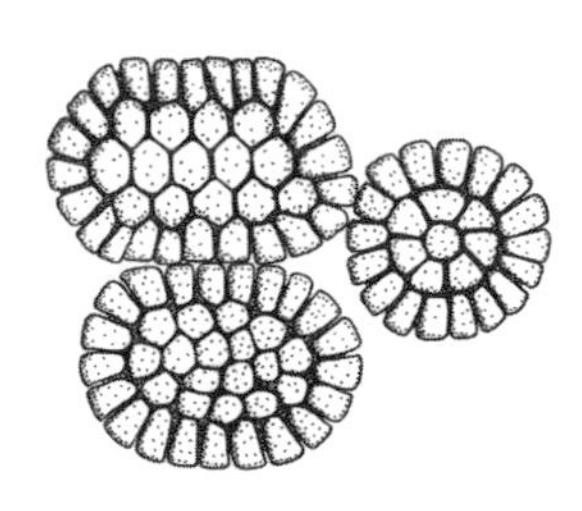

61. Aboral tabulate plates of *M. aequalis.*

Mediaster aequalis is a flat, broad-armed sea star with five arms up to 10 cm long. It is vermilion aborally and more orange orally, with red to flesh-coloured tube feet. The ratio of arm to disc is from 2.4 to 3.0. The aboral tabulate plates are hexagonal to round and bear up to 24 flat-topped, roundish central granules and up to 25 slightly longer peripheral granules (figure 61). Two or three papulae occur between the plates, except in a triangular area in the interradial region where there are none. *M. aequalis* has 28 conspicuous superomarginals that are not bulging or separated from the inferomarginals by an obvious groove, as in *Gephyreaster swifti*. The inferomarginals are similar to the superomarginals, and there is a slight zigzag line between the two series (figure 62). The oral interradial area is extensive, with oral intermediates nearly to tip of the arm, each one oval, with roundish groups of 2 or 3 central granules and 5 to 10 peripheral. The adambulacrals are squarish with 3 to 5 blunt prismatic furrow spinelets and two longitudinal series of 3 shorter blunt spines on the oral surface (figure 63). The mouth plates have a marginal series of 5 to 7 prismatic spinelets, and 3 to 5 thicker prismatic suboral spinelets and granules on the outer part of the plate.

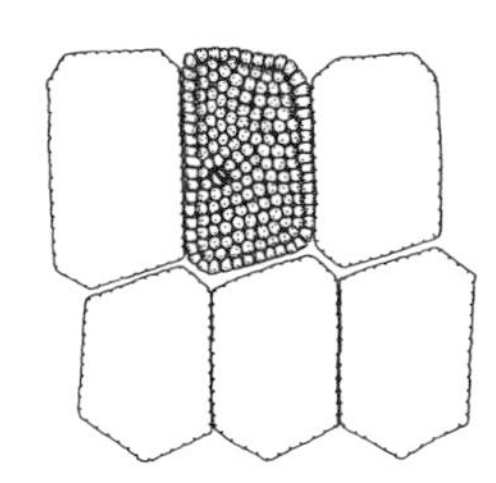

62. Marginal plates (superomarginals above and inferomarginals below).

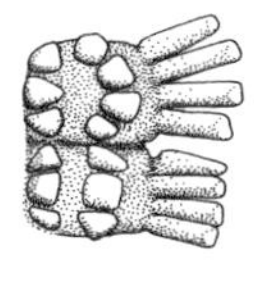

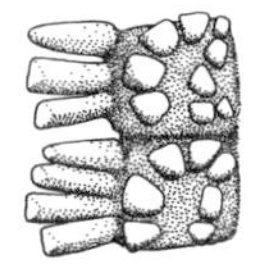

63. Adambulacral spines.

Similar Species

Mediaster aequalis may be confused with *Gephyreaster swifti* or *Pseudarchaster alascensis*, but it differs mainly in the form and arrangement of the marginal plates. Compare the descriptions of marginal plates for the three species.

Distribution

Chignik Bay, Alaska Peninsula, to southern California; intertidal to 293 metres. Common in shallow subtidal waters on rocks, shells, sand, gravel, pebbles and mud.

Biology

The diet of *Mediaster aequalis* varies with the substrate and season. It eats encrusting sponges, bryozoans, the sea pen *Ptilosarcus*, loose algae, detritus and dead animals. In Gabriola Passage in December, 41 per cent were feeding and, of those, 56 per cent ate detritus.

M. aequalis reaches sexual maturity in about four years. It breeds from March to May. A specimen with a radius of 65 mm can produce about 1,800 eggs per year. The eggs, 1.0 to 1.2 mm in diameter, are bright opaque orange. The larvae are lecithotrophic. In the laboratory, the presence of the tubes of the polychaete worm *Phyllochaetopterus* caused larvae to settle to bottom after 30 days. In nature, these worm tubes are a nursery area for juvenile *M. aequalis* as well as *Crossaster papposus, Luidia foliolata, Pteraster tesselatus, Henricia leviuscula, Solaster stimpsoni* and *S. dawsoni*. *M. aequalis* can move at speeds of 27 to 40 cm per minute.

References

Birkeland et al. 1971, Carey 1972, Hagiwara 1979, Hopkins and Crozier 1966, Lansman 1983a, b and c, Mauzey et al. 1968, McEdward and Chia 1991, Moody and Hagiwara 1982, Sloan and Robinson 1983, Webster 1975. **Range:** Fisher 1911.

Pseudarchaster alascensis

alascensis: **from the type locality in Alaska.**

Description

Pseudarchaster alascensis has marginal plates in a continuous band around the perimeter of the aboral surface, and the joints between plates are ill-defined. It is reddish-orange aborally and paler on the oral side. Its five arms can grow up to 10 cm long. The ratio of arm to disc is from 2.6 to 4.0. The aboral tabulate plates are small and hexagonal, forming a relatively smooth surface. Each plate is topped with 15 to 22 granules and has 6 papulae around it. *P. alascensis* has 33 to 58 superomarginals that encroach well onto the aboral surface, but the individual plates are not well defined. Compact hexagonal granules cover each plate. The corresponding inferomarginals have a few broad, pointed spinelets, as well as granules. The oral intermediates form a triangular area comprised of rows of plates connecting between adambulacrals and inferomarginals up to the 10th or 17th inferomarginal. Each plate is armed with unequal, crowded granules, the central ones flat and lance-shaped, as on inferomarginals. The adambulacrals bear 5 or 6 flat furrow spines, and some shorter, stouter spines continue around the edge of each plate. The oral surface has 1 to 3 more robust, bluntly pointed spines. The adjoining mouth plates share a single large apical spine, and each has about 7 marginal spines.

64. *Pseudarchaster alascensis*, R = 52 mm. (See colour photograph C-12.)

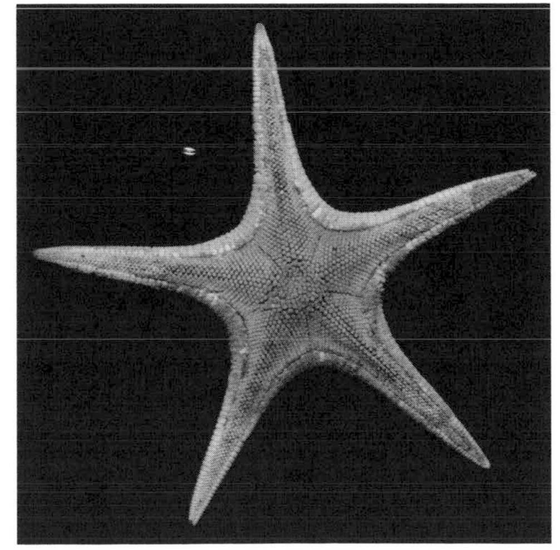

Taxonomic Note: Originally described as *Pseudarchaster parelii alascensis* Fisher; revised to *Pseudarchaster alascensis* by Halpern (1972).

Similar Species

Pseudarchaster alascensis might be confused with *Mediaster aequalis* or *Gephyreaster swifti*. Compare the descriptions and illustrations of the marginal plates in these species.

Distribution

Sea of Japan, Sea of Okhotsk, and Bering Sea to Oregon in depths of 92 to 1947 metres. Common on mud, sand, gravel and rock, but most often on soft substrates. *Pseudarchaster parelii* (Duben and Koren, 1846) is considered circumboreal at depths of 15 to 2,500 metres, but Halpern (1972) proposed that the subspecies *P. parelii alascensis* Fisher, 1911, be raised to species status.

Biology

The stomachs of *Pseudarchaster alascensis* specimens contained gastropods, brittle stars, small crustaceans, foraminiferans and sediment. This sea star appears to reach full size in four years. Its breeding season is not known.

References

Carey 1972, Halpern 1972. **Range:** Alton 1966, Fisher 1911.

Family Asteropseidae

Characteristics: Large disc (usually); short, broad arms. The aboral skeletal plates are sometimes meshlike, in regular radial rows, but usually covered with thick skin, which can be smooth, granulated or spiny. The marginal plates can vary from prominent to poorly developed, usually oblique or overlapping, sometimes with a small group of spinules on the outer edge or with a single spine; the edge of the disc is thin. The oral intermediates are parallel to the adambulacrals. The adambulacral plates have two series of spines sheathed in thick skin. The papulae can be single or grouped in dorsal radial areas. Bivalved or granuliform pedicellariae are sometimes present.

In this region:
Dermasterias imbricata.

Dermasterias imbricata **Leather Star**

imbricata: **Latin *imbricatus*, overlapping, referring to the arrangement of aboral plates.**

Description

Dermasterias imbricata has a characteristic smooth, slimy feel and smells like garlic. It is mottled reddish-brown and orange with a greenish-grey perimeter and rusty brown gill areas. The madreporite stands out as a large yellow dot. The five arms are up to 15 cm long and the arm-to-disc ratio is from 2.4 to 2.9. Few if any calcareous plates show through the leathery skin. Along each arm are six to eight rows of squarish areas containing up to 22 papulae and 1 to 9 pedicellariae in each (figure 66). The smooth shape of the marginals shows through the leathery skin especially in a dried specimen. Each adambulacral has a single blunt furrow spine tipped with a fleshy knob and on the oral surface, a ridge of tough skin parallel to the furrow and a flat oval spinelet. *D. imbricata* has four rows of tube feet.

Taxonomic Note: Originally described as *Asteropsis imbricata* Grube; revised to *Dermasterias imbricata* by Sladen (1889).

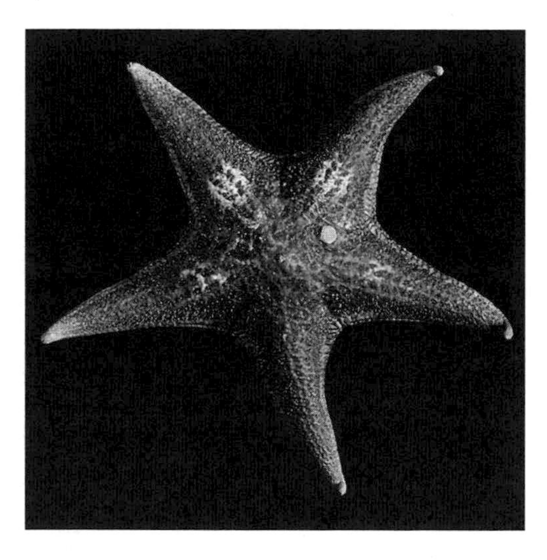

65. *Dermasterias imbricata*, R = 120 mm. (See colour photograph C-13.)

Similar Species

Dermasterias imbricata might be confused with *Asterina miniata* in general shape, but *A. miniata* is stiffer and feels like sandpaper.

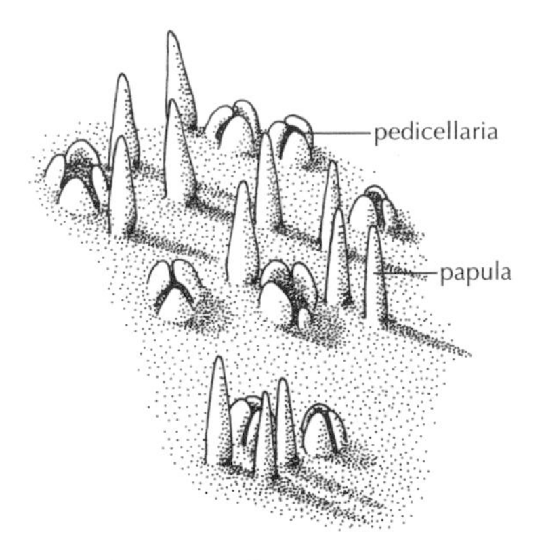

66. Aboral surface.

Distribution

Cook Inlet, Alaska, to La Jolla, California, from the intertidal zone to 91 metres depth on rocky substrates. Common in this region.

Biology

The diet of *Dermasterias imbricata* varies with locality. On the exposed coast, it eats primarily sea anemones and some compound sea squirts, and in sheltered waters, mostly sea cucumbers as well as encrusting sponges and sea pens. Even though the sea cucumber *Psolus chitonoides* contains toxic saponins, 39 per cent of *D. imbricata* surveyed were feeding on it. All those observed in Gabriola Passage were feeding on bottom detritus. This sea star evokes a swimming-escape response in the sea anemones *Stomphia didemon* and *S. coccinea*, and presumably would eat them if it was able to catch them. *Stomphia* species detect imbricatine, an alkaloid chemical produced by *D. imbricata*. Seven other species of sea stars cause *Stomphia* species to release but do not eat them. *D. imbricata* also stimulates juveniles of the anemone *Urticina piscivora* to release their attachment; large *U. piscivora* may defend themselves by attacking the sea star, though they later regurgitate it. In California, *D. imbricata* usually feeds on the anemone *Corynactis californica*. In an aquarium, it prefers the sea anemones *Anthopleura* or *Metridium*; but in the ocean, these anemones are not regular prey items because in the intertidal zone they live too high and in the subtidal zone they grow too large for the sea star.

D. imbricata breeds from April to August. It sheds gametes via three paired gonopores in each arm radius. The yellow-to-orange eggs are 200 micrometres in diameter. In the Vancouver Island region, juveniles successfully settled and survived just once in five years. The scale worm *Arctonöe vittata* is found on *D. imbricata*; a study at Bodega Bay discovered that the sea star was attracted to the scale worm more than to food items, suggesting that the relationship may be of mutual benefit. The parasitic barnacle *Dendrogaster* can infect *D. imbricata*'s reproductive organs.

References

Annett and Pierotti 1984, Bakus 1974, Bavendam 1985, Bigger and Hildemann 1982, Bingham and Braithwaite 1986, Birkeland et al. 1982, Bruno et al. 1990 and 1992, Clark 1984, Cool et al. 1988, Dalby et al. 1988, Elliott et al. 1985, Elliott et al 1989, Estell and Laskowski 1980, Estell et al. 1980, Francour 1997, Hart 1991, Hopkins and Crozier 1966, Kaneshiro and Karp 1980, Mauzey et al. 1968, Pathirana and Andersen 1986, Sewell and Watson 1993, Sloan and Robinson 1983, Smith 1982, Smith et al. 1980, Smith et al. 1982, Stickle 1985c, Van Veldhuizen and Oakes 1981, Wagner et al. 1979, Ward 1965a and 1965b, Webster 1975. **Range:** Barr and Barr 1983, Morris, et al. 1980.

ORDER VELATIDA

Family Solasteridae

Characteristics: Broad disc with five or more long arms. The aboral skeleton is meshlike; the plates bear pseudopaxillae. The marginal pseudopaxillae are larger than the aborals. Oral intermediates are present. The adambulacral spines consist of two series at right angles to each other. The mouth plates are prominent. No pedicellariae.

In this region:
Crossaster papposus
Lophaster furcilliger vexator
Solaster dawsoni
Solaster endeca
Solaster paxillatus
Solaster stimpsoni

Deeper than 200 metres:
Lophaster furcilliger
Solaster borealis

Crossaster papposus **Rose Star, Snowflake Star**

papposus: **Latin *pappus*, bristles, teeth, referring to the spiny surface.**

Description

Crossaster papposus is a colourful sea star with 8 to 16 arms, usually 10 or 11, up to 8.5 cm long. Its colour is varies greatly: it may be solid purple or red, but it usually has a concentric pattern of bright red, orange, white or yellow on its spiny aboral surface. The ratio of arm to disc ranges from 1.8 to 2.7. The aboral pseudopaxillae are widely spaced and bear up to 50 slender spinelets. The marginals consist of a single series of prominent pseudopaxillae. The oral interradial area has up to about 25 pseudopaxillae. The adambulacrals usually have 3 to 5 furrow spinelets with a transverse comb of 5 to 9 slightly longer stouter spinelets on the oral surface (figure 67). The mouth plates have 8 to 10 marginal spines with 2 to 4 suborals.

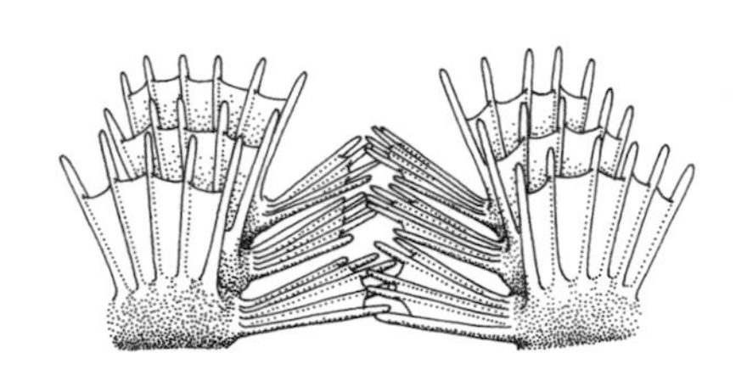

67. Adambulacral spines of *C. papposus.*

Taxonomic Note: Originally described as *Asterias papposa* Linnaeus; also known as *Solaster papposus* (Linnaeus); revised to *Crossaster papposus* by Müller and Troschel (1840).

Similar Species

Crossaster papposus might be mistaken for a juvenile *Pycnopodia helianthoides*. But *P. helianthoides* is generally much larger, has more arms, differs in the detail of the aboral surface, adambulacral and mouth-plate spines and has four rows of tube feet.

Distribution

Circumpolar. In the Pacific, to Washington and the Sea of Okhotsk; in the Atlantic, to 40°N latitude on the American side, and to Scandinavia and the British Isles on the northern European coast. Found from the intertidal zone to 1200 metres deep on soft mud, gravel, sand, pebbles or rock. Common in this region.

68. *Crossaster papposus*, R = 63 mm. (See colour photograph C-14.)

Biology

In British Columbia, *Crossaster papposus* eats the sea pen *Ptilosarcus*, sea slugs (opisthobranchs), bryozoa, sea squirts and bivalves. It may also attack other sea stars (such as *Evasterias troschelli*). In Scotland, *C. papposus* breeds from March to April; each sea star may spawn several times at intervals of two to ten days, shedding more than 2000 eggs at a time, 6000 altogether. The eggs (0.8 mm) are reddish-brown to clay colour and are fertilized externally. The lecithotrophic larva settles to the bottom after 18 days and attaches by a sucker. The mouth forms after about 38 days. Juveniles are often found among the tubes of the polychaete worm *Phyllochaetopterus*. *C. papposus* moves at a rate of 60 to 70 cm per minute. A well-fed Rose Star grew from 40 mm to 90 mm in one year. Richard Carlson at the Auke Bay Lab near Juneau studied a population for 17 years. He found that growth rates were quite slow, with individuals achieving a maximum diameter of 30 cm in about 10 years. Their principal prey item was the scallop *Chlamys rubida*. *Crossaster papposus* can live for at least 20 years. Mortality rates were low and recruitment of young was infrequent in that population. The scale worm *Arctonöe vittata* is commensal on *C. papposus*.

References

Birkeland et al. 1971, Carlson and Pfister 1999, Dembetskii 1988, Francour 1997, Gemmill 1920, Himmelman 1991, Himmelman and Dutil 1991, Holme 1966, Lafay et al. 1995, Legault and Himmelman 1993, Levin et al. 1984, Mauzey et al. 1968, Sloan 1977, 1979, 1980 and 1984, Sloan and Northway 1982. **Range:** D'Yakonov 1950, Fisher 1911.

69. *Lophaster furcilliger vexator*, R = 67 mm. (See colour photograph C-15.)

Lophaster furcilliger vexator

furcilliger: **Latin *furcilla*, fork, referring to the pseudopaxillae with forked spines.**
vexator: **Latin *vexatus*, annoy, perhaps because Fisher had so much trouble identifying this subspecies.**

Description

Lophaster furcilliger vexator is pastel yellow with orange on the arms; the dark brown pyloric caeca shows through skin of the aboral surface. It has five arms up to 8.2 cm long. The ratio of arm to disc is from 3.3 to 4.4. The aboral surface consists of a coarse meshwork of calcareous skeleton with a stout pseudopaxilla arising at the junction of each mesh. Each pseudopaxilla has 20 or more glassy spinelets with two to four delicate teeth at the tip. The papulae form groups of 2 to 12 (figure 70, left). Approximately 24 well-spaced conspicuous pseudopaxillae occur along supero- and inferomarginals (figure 70, right). The oral interradial area is normally triangular with 8 or 9 pseudopaxillae. Oral intermediates form a single row from the third inferomarginal to the tip of the arm (figure 70, right). The adambulacrals are short and wide, the spaces between the plates about equal to the length; each has five furrow spinelets, sometimes reduced to one at the tip of the arm. On the oral surface, they have a transverse series of 2 to 4 longer, more robust spinelets. The mouth plates have 7 or 8 marginal spinelets joined by a web with 7 or 8 suborals along the suture.

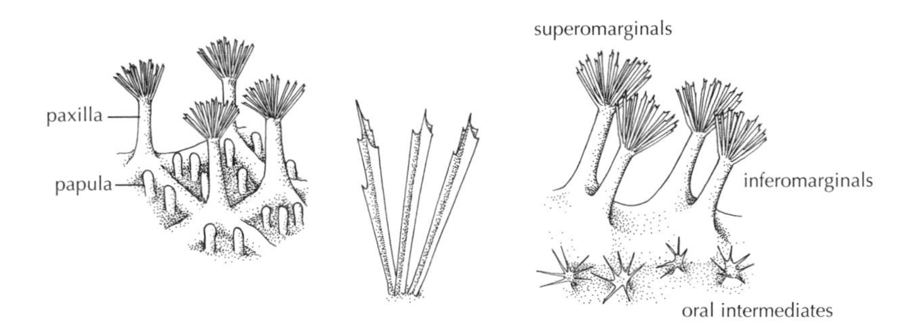

70. Identifying features of *L.f. vexator,* left to right: aboral surface, pseudopaxillar spines and marginal spines.

Taxonomic Note: *Lophaster furcilliger* occupies deeper water than *L.f. vexator* and has a smaller disc, thinner arms and more slender pseudopaxillae. Its arm-to-disc ratio is 4.4. It seems to intergrade with *L.f. vexator,* which Fisher (1911) suspected was intermediate to the Atlantic species *Lophaster furcifer* but chose to keep it separate. Grieg (1921) maintains that specimens of *L. furcifer* in his possession show transitional forms between *L. furcilliger vexator* and *L. furcifer*. Although a close relationship among these three forms has been suggested by several authors, no one has yet designated them as the same species.

Similar Species

The only other species in this book with similar surface detail to *Lophaster furcilliger vexator* is *Crossaster papposus,* but the colour and number of arms separate them easily.

Distribution

Lophaster furcilliger vexator occupies shallow water and ranges from the southern Bering Sea to northern California, from 21 to 670 metres depth, but usually less than 360 metres. The deep-water form, *Lophaster furcilliger,* is recorded from south of the Alaska Peninsula to southern California and to the Galapagos Islands at depths of 350 to 2010 metres. *L. f. vexator* is commonly dredged on mud; less often it is found by scuba divers on rocky substrates in less than 30 metres.

Biology

Stomachs of *Lophaster furcilliger vexator* specimens from soft substrates contained the remains of sea urchins, brittle stars, tube-dwelling polychaetes, foraminifera and detritus. Nothing is known about the feeding habits of specimens from shallow rocky substrates.

References

Anderson and Shimek 1993, Carey 1972, Fisher 1911, Grieg 1921, McDaniel et al. 1978. **Range:** Fisher 1911, Lambert 1978b.

Solaster dawsoni **Morning Sun Star**

dawsoni: **after Dr G.M. Dawson, Geological Survey of Canada.**

Description

Solaster dawsoni is usually a homogeneous brown, with an evenly textured aboral surface. It has 11 or 12 arms (sometimes 8 to 13) and can grow up to 20 cm in radius. The ratio of arm to disc is from 2.5 to 3.4. Its body is occasionally red, orange, or mottled brown and beige. When viewed from above, the inferomarginals appear as serrated edges of the arms. The aboral pseudopaxillae are round and flat-topped with 30 tiny spinelets on the perimeter and up to 15 central spinelets. The superomarginals are similar in shape but larger. The oblong inferomarginals are three to four times wider than their length. The oral interradial area is small (only 20 to 25 plates) when compared with the three other species in this genus. The adambulacrals have a furrow series of 3 or 4 long, tapering spines; the oral surface bears a straight transverse series of 3 to 6 long spines that stand about twice as high as the spines on the adjoining inferomarginals and slightly higher than the furrow series (figure 72). Each pair of mouth plates is rather broad and shovel-like with 9 to 11 marginal spines, the apical spine being

71. *Solaster dawsoni*, R = 145 mm. (See colour photograph C-16.)

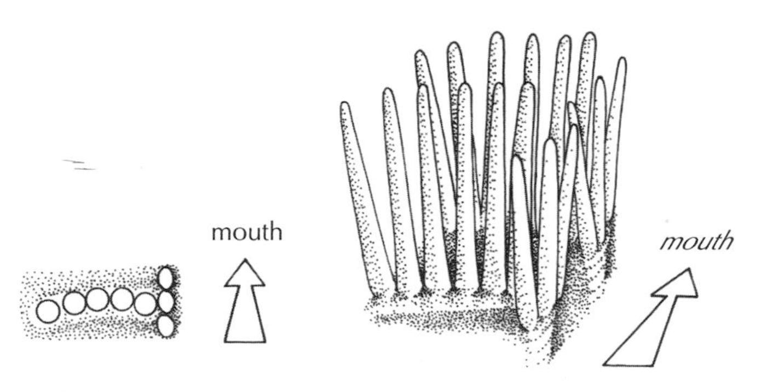

72. Adambulacral spines of *S. dawsoni*, top view (left) and oblique view (right).

largest. Spines on the oral surface can vary, from 1 or 2 suborals to a double row of 5 to10 spines.

Similar Species

The living colour and body shape of the four species of *Solaster* in British Columbia are fairly distinctive, but if specimens are dried you may need to consult the descriptions of the adambulacral spines.

Distribution

From Point Franklin on the north side of Alaska to Monterey Bay, California, and on the Asiatic side to the Yellow Sea. Found from the intertidal zone to 420 metres deep on rocky substrates. Uncommon.

Biology

Solaster dawsoni is carnivorous. In 36 of 55 feeding observations in Puget Sound, it preyed on *S. stimpsoni* and consumed the sea star in an average of 4.5 days. This predator ate about 17 *S. stimpsoni* per year. In the Strait of Georgia, *Dermasterias imbricata* seems to be the prey of choice based on diver observations. *S. dawsoni* also attacks its own species, but its success rate is low because of a well-developed escape response. *S. dawsoni* bends its arms back and pushes the attacker off while rapidly moving away (10 cm/min.). This sea star also preys upon *Crossaster papposus, Leptasterias hexactis, Mediaster aequalis, Evasterias troschelii, Henricia leviuscula,*

Asterina miniata, small *Pisaster ochraceus* and *Pycnopodia helianthoides*. When attacking, *S. dawsoni* can reach a speed of 12 cm per minute. In a Bodega Bay study, *D. imbricata*, *Pisaster brevispinus* and large *P. ochraceus* appeared to avoid predation; the latter two used their pedicellaria to ward off *S. dawsoni*. The strongest escape responses are shown by *A. miniata*, *H. leviuscula*, *L. hexactis* and *Pycnopodia helianthoides*; and the California Sea Cucumber (*Parastichopus californicus*) swims away when contacted.

S. dawsoni breeds from late February to early March. Large buoyant eggs (0.94-1.05 mm) undergo total and equal cleavage to a planktonic lecithotrophic larva with three brachiolarian arms. Settlement occurs 8-10 days after fertilization and metamorphosis 40-50 days after. The juvenile has five arms initially and adds the remaining arms in sequence between the first and fifth. I observed three specimens spawning on April 13 1977 near Victoria. Juveniles have been reported to be quite common among tubes of the polychaete *Phyllochaetopterus*. *S. dawsoni* has two commensal scale worms: *Arctonöe fragilis* and *A. vittata*.

References

Birkeland et al. 1971, Birkeland et al. 1982, Francour 1997, Margolin 1976, Mauzey et al. 1968, McDaniel 1977, McEdward and Chia 1991, Nance and Braithwaite 1979, Van Veldhuizen and Oakes 1981. **Range:** D'Yakonov 1950, Fisher 1911.

Solaster endeca Northern Sun Star

Description

Solaster endeca has a broad disc and 7 to 13 – usually 9 to 11 – short, sharply tapering arms. It is usually red or orange aborally and yellow or beige on oral side, but it occasionally has a purple stripe down each arm, similar to *S. stimpsoni*. The arm length can be up to 20 cm; the average length in British Columbia is 9 cm. The ratio of arm to disc is from 2.3 to 3.3. *S. endeca* has numerous fine, crowded aboral pseudopaxillae and 5 to 9 spinelets with single papulae between. Its disc is thick and tapers down to the arm tips. The superomarginals are only slightly larger than the aboral pseudopaxillae, but the wide inferomarginals alternate with them and form a very obvious series. The size of oral interradial area varies with the number of arms but is larger than the other species of *Solaster*. The adambulacral spines do not protrude noticeably above the general oral surface. There are 1 to 4 furrow spines deep in the furrow and a curved transverse series of 6 to 8 sharp spines on the oral surface. The inner and outermost spines of the curve are farthest from the mouth (figure 74). The mouth plates have 7 to 9 marginal spines (the apical spines are largest) and 6 to 15 suborals in two series or in a triangular group.

Taxonomic Note: Originally described as *Asterias endeca* Linnaeus; revised to *Solaster endeca* by Forbes (1839). The derivation of the species name *endeca* is unknown.

73. *Solaster endeca*, R = 85 mm. (See colour photograph C-17.)

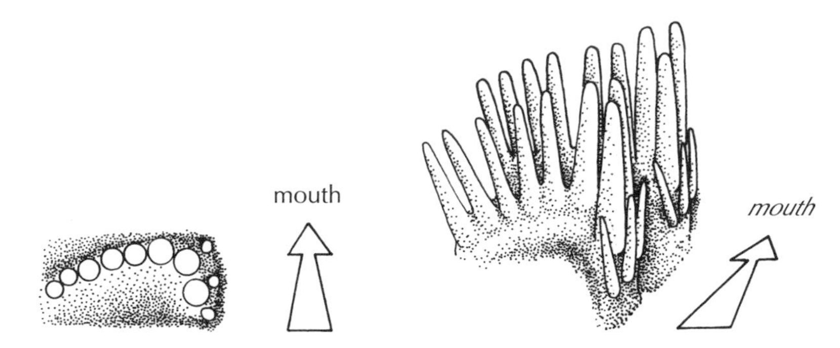

74. Adambulacral spines of *S. endeca,* top view (left) and oblique view (right).

Similar Species

Solaster endeca differs from *S. stimpsoni* in its arm-to-disc ratio and the arrangement of the adambulacral spines.

Distribution

Circumboreal: Arctic Ocean; in the North Atlantic to Great Britain and Cape Cod; in the North Pacific to Puget Sound. Found from the intertidal zone to 475 metres deep. In shallow water, it is usually on rocky substrates, but in the deeper part of its range, it is often on pebbles or mud. Uncommon.

Biology

In the Atlantic, *Solaster endeca* is reported to be a voracious feeder on sea stars and molluscs, but on the Pacific coast it neither attacks nor evokes escape responses in other sea stars. Of 29 specimens examined, 21 were feeding on the small sea cucumber *Cucumaria lubrica*, 2 on *C. miniata* (Orange Sea Cucumber), and the remainder on bryozoans, sea squirts, and unidentified organisms. *S. endeca* causes the California Sea Cucumber (*Parastichopus californicus*) to swim when contacted. Near Juneau, Dr Rita O'Clair reported that *S. endeca* fed on the False White Sea Cucumber (*Eupentacta pseudoquinquesemita*) in the intertidal zone.

S. endeca breeds from March to April. Its large yolky eggs (1 mm in diameter) develop into non-feeding planktonic larvae. After about 20 days, they form small tube feet and settle to the bottom. Unlike *Pycnopodia helianthoides* and *S. dawsoni*, in British Columbia *S. endeca* does not elicit escape responses in the abalone *Haliotis*, key-hole limpet *Diodora*, turban snail *Tegula* or the swimming scallops *Chlamys* spp. The scale worm *Arctonöe vittata* is commensal on *S. endeca*.

References

Birkeland et al. 1982, Burnell et al. 1986, Dembetskii 1988, Francour 1997, Gemmill 1912, Himmelman 1991, Himmelman and Dutil 1991, Margolin 1976, Mauzey et al. 1968, McDaniel 1977, McEdward and Chia 1991, Ray et al. 1980. **Range:** D'Yakonov 1950.

75. *Solaster paxillatus*, R = 175 mm. (See colour photograph C-19.)

Solaster paxillatus **Orange Sun Star**

paxillatus: **Latin *paxillus*, peg, small stake.**

Description

Solaster paxillatus has 8 to 10 arms up to 19 cm long. The arms are broad at the base, narrowing rapidly to the tip. The body is orange aborally and yellow orally. A specimen off Oregon at 457 metres was violet on both sides. Viewed from above, the marginals create a serrated edge, just as in *S. dawsoni*, but the arrangement of the adambulacral spines will separate these two species. The ratio of arm to disc is from 3.0 to 3.7. The aboral pseudopaxillae are evenly spaced and each bears a crown of numerous fine spinelets about as long as the pseudopaxilla is high. Between each arm there is often a suggestion of a bare streak. On dried specimens the aboral skeleton forms a squarish or diamond-shaped mesh with a pseudopaxilla at each junction of the mesh. The papulae occur in groups of 1 to 12. About 60 large fan-shaped inferomarginals create a serrated edge to the arm. Superomarginals are about the same size as the largest aboral pseudopaxilla and situated just above and between each inferomarginal. The oral interradial area contains numerous pseudopaxillae topped with a compact cluster of spinelets slightly longer than the base of the pseudopaxilla. The adambulacrals are similar in general appearance to those of *S. dawsoni*, with 4 furrow

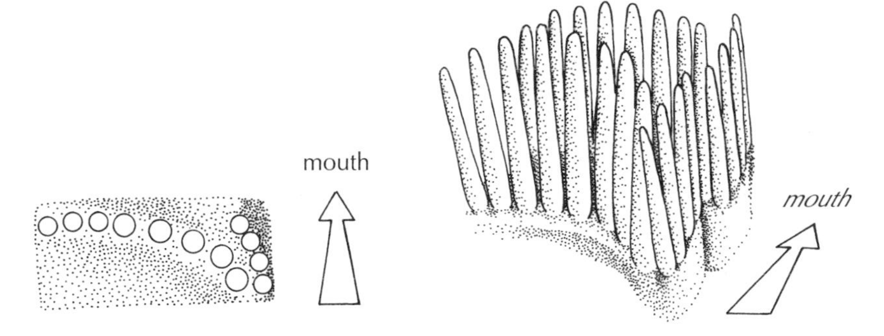

76. Adambulacral spines of *S. paxillatus*, top view (left) and oblique view (right).

spines reducing to 1 distally, and on the oral surface of the plate, a curved transverse series of 5 to 8 long slender spines only slightly longer than the furrow spines. The proximal spine of the transverse series meets the distal spine of the furrow series (figure 76). The mouth plates are large and shovel-like with 8 to 11 webbed marginal spines and up to 20 finer suboral spines.

Similar Species

Solaster paxillatus differs from other *Solaster* species in body proportions and adambulacral spines.

Distribution

The Bering Sea to northern Oregon and to Yokohama, Japan. Specimens have been collected in depths of 11 to 640 metres on rock, gravel and mud. Rare in this region. Three specimens recorded in British Columbia and one each in Washington and Oregon.

Biology

Solaster paxillatus is probably carnivorous, like other species of *Solaster*. Stomach contents recorded: the Sweet Potato Sea Cucumber (*Molpadia intermedia*) and the Blood Star (*Henricia leviuscula leviuscula*).

References

Fisher 1911, Lambert 1978a. **Range:** Lambert 1978b.

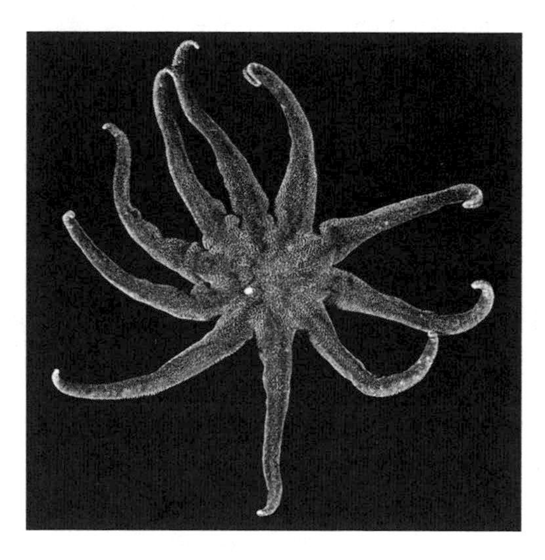

77. *Solaster stimpsoni*, R = 225 mm. (See the colour photographs C-18 and on the front cover.)

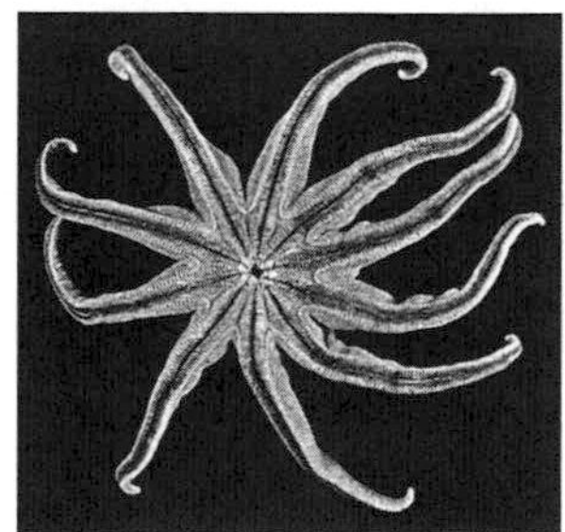

Solaster stimpsoni **Striped Sun Star**

stimpsoni: **after Dr William Stimpson, Smithsonian Institution, 1850s.**

Description

Solaster stimpsoni is the most common of the four species of *Solaster*. It has ten arms (occasionally nine) up to 23 cm long. The body is typically orange or yellow with a blue or purple band on the aboral side of each arm that join at the central disc. Less commonly, the whole body is blue. The ratio of arm to disc is from 2.3 to 4.4. The aboral pseudopaxillae are intermediate in size and number between *S. endeca* and *S. dawsoni*, with 6 to 12 peripheral spinelets per pseudopaxilla. The marginals are confined to the oral side of the arms; the wide inferomarginals, at right angles to the arm axis, form an obvious regular row adjacent to the adambulacrals. The superomarginals, just above and between each inferomarginal, are about the same size as the aboral pseudopaxillae. The oral interradial area has pseudopaxillae similar to those on aboral side. A row of oral intermediates runs about one-fifth of the way along the arm. The adambulacrals have 2 or 3 short furrow spines, half as long as the transverse series of 6 to 8 blunt spines on the oral surface (figure 78). Each pair of mouth plates is broad and shovel-shaped with 6 to 8 heavy marginal spines, larger than any others on the sea star, and one row of 4 to 8 suborals.

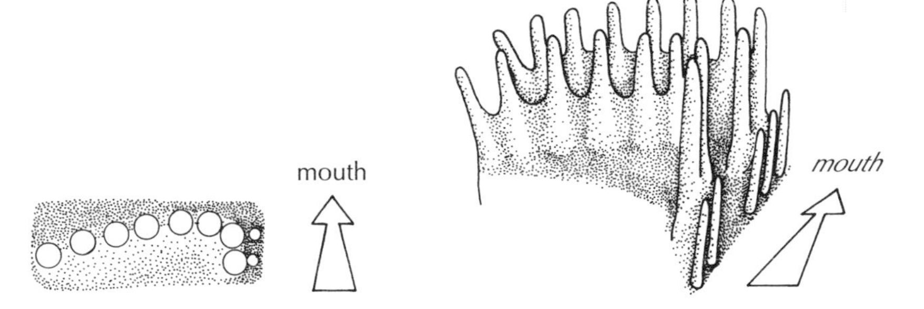

78. Adambulacral spines of *S. stimpsoni*, top view (left) and oblique view (right).

Similar Species

Solaster endeca may have a striped arm as well, but the arm-to-disc ratio is much smaller than *S. stimpsoni.*

Distribution

The south Bering Sea to Oregon and to Japan, from the intertidal zone to 60 metres deep. Usually on rocky substrates; less commonly on sandy mud. Common in this region.

Biology

In Puget Sound, *Solaster stimpsoni* feeds primarily on *Pseudocnus lubricus* (the Aggregating Sea Cucumber), and less commonly on these other sea cucumbers: *Cucumaria miniata, Eupentacta quinquesemita, E. pseudoquinquesemita* and *Psolus chitonoides.* It seems to prefer the *Eupentacta* species, even though it can find *P. lubricus* in large aggregations. *Parastichopus californicus* (the California Sea Cucumber) swims away when contacted by *S. stimpsoni.* A small number feed on sea squirts, lamp shells and sea pens. *S. stimpsoni* is not known to feed on other sea stars.

This sea star breeds from late February to early March. Its large, buoyant eggs (0.94-1.05 mm in diameter) have total and equal cleavage and develop into a planktonic lecithotrophic larva with three brachiolarian arms. The larva settles 8-10 days after fertilization. It forms five arms initially, then adds more in sequence between the first and fifth arms. Metamorphosis is complete 40-50 days after fertilization. Juveniles are often found among the tubes of the polychaete worm *Phyllochaetopterus. S. stimpsoni* has two commensal scale worms: *Arctonöe pulchra* and *A. vittata.* Twenty-one per cent of specimens sampled were infected with a parasitic green alga, *Diogenes.*

References

Birkeland et al. 1971, Birkeland et al. 1982, Carson 1988, Engstrom 1988, Francour 1997, Mauzey et al. 1968, McDaniel 1977, McEdward and Carson 1987, Smith 1982, Smith et al. 1982.
Range: D'Yakonov 1950.

COLOUR PHOTOGRAPHS

C-0. A Blood Star (*Henricia leviuscula leviuscula*) at Macaulay Point, Victoria. (See also C-23.)

C-1. Bat Stars (*Asterina miniata*) at Klaskino Inlet on the west coast of Vancouver Island. Page 50.

C-2. A Sand Star (*Luidia foliolata*) dredged from a depth of 60 metres off Calvert Island, on the central coast of B.C. Page 35.

C-3. *Leptychaster pacificus* dredged from Howe Sound, just north of Vancouver. Page 42.

C-4. A Mud Star (*Ctenodiscus crispatus*) dredged from 270 metres deep in Howe Sound, just north of Vancouver. Page 44.

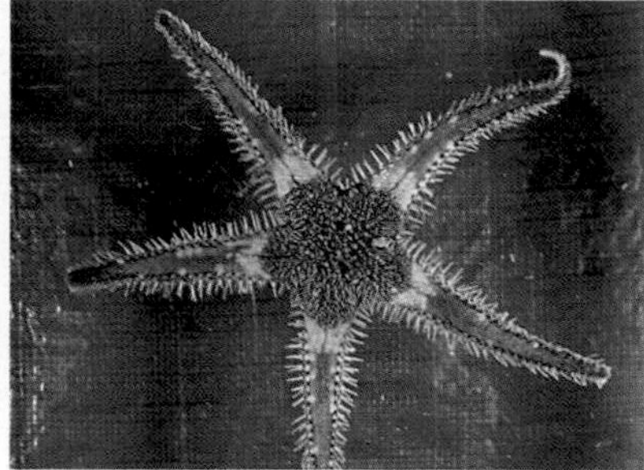

C-5. *Cheiraster dawsoni* trawled from a depth of 235 metres near Rose Spit in Dixon Entrance. Page 48.

C-6. *Poraniopsis inflatus inflatus* beside a Red Sea Urchin (*Strongylocentrotus franciscanus*) in Shields Bay, Queen Charlotte Islands. Page 54.

C-7. A Gunpowder Star (*Gephyreaster swifti*) near Nigei Island off northeastern Vancouver Island. Page 62.

C-8. An Arctic Cookie Star (*Ceramaster arcticus*) at Race Rocks, Juan de Fuca Strait. Page 58.

C-9. A Cookie Star (*Ceramaster patagonicus*) in Desolation Sound, northern Strait of Georgia. Page 60.

C-10. A Spiny Red Sea Star (*Hippasteria spinosa*) in Barkley Sound. Page 64.

C-11. A Vermilion Star (*Mediaster aequalis*) at Donaldson Island, Juan de Fuca Strait. Page 67.

C-12. *Pseudarchaster alascensis* trawled from 270 metres deep in Howe Sound. Page 69.

C-13. A Leather Star (*Dermasterias imbricata*) at Sitka, Alaska. Page 72.

C-14. A Rose Star (*Crossaster papposus*) in Barkley Sound. Page 76.

C-15. *Lophaster furcilliger vexator* at a depth of 20 metres near Edward King Island, Barkley Sound. Page 79.

C-16. A Morning Sun Star (*Solaster dawsoni*) in Saanich Inlet, near Victoria, B.C. Page 81.

C-17. A Northern Sun Star (*Solaster endeca*) near Taylor Islet, Barkley Sound. Page 84.

C-18. A Striped Sun Star (*Solaster stimpsoni*) in the waters off Rocky Point, west of Victoria, B.C. Page 89.

C-19. An Orange Sun Star (*Solaster paxillatus*) at Beaver Harbour, near Port Hardy, northeastern Vancouver Island. Page 87.

C-20. A Wrinkled Star (*Pteraster militaris*) off Cabbage Island, Gulf Islands, southern Strait of Georgia. Page 94.

C-21. A Cushion Star (*Pteraster tesselatus*) off Edward King Island, Barkley Sound. Page 96.

C-22. A Ridged Blood Star (*Henricia aspera aspera*) in Tasu Sound, Queen Charlotte Islands. Page 100.

C-23. A Blood Star (*Henricia leviuscula leviuscula*) at Brothers Islets, near Victoria, B.C. Page 104.

C-24. A Fat Henricia (*Henricia sanguinolenta*) draped over a Boot Sponge (*Rhabdocalyptus dawsoni*), near Sonora Island, northern Strait of Georgia. Page 109.

C-25. An undescribed species of *Henricia* found under rocks in the intertidal zone (page 105). This one was seen at Ten Mile Point, Victoria, B.C.

C-26. Three colour variations of the Mottled Star (*Evasterias troschelii*) at Lena Point, Lynn Canal, just south of Skagway, Alaska. Page 113.

C-27. The Six-armed Star *Leptasterias aequalis* at Race Rocks, Juan de Fuca Strait. Page 118.

C-28. The Six-armed Star *Leptasterias alaskensis* in the intertidal zone at Port Renfrew, east of Victoria. It is regenerating two of its arms. Page 119.

C-29. The Six-armed Star *Leptasterias hexactis* at Sitka, Alaska. Page 120.

C-30. A preserved specimen of *Leptasterias coei,* collected from 10 metres in Gastineau Channel, near Juneau, Alaska. RBCM 992-95-1. Page 123.

C-31. A preserved specimen of *Leptasterias polaris katherinae* in the Auke Bay Laboratory's collection (AB90-20). Page 125.

C-32. *Lethasterias nanimensis* in Lynn Canal, near Juneau, Alaska. Page 127.

C-33. A Rainbow Star (*Orthasterias koehleri*) at Race Rocks, Juan de Fuca Strait. Page 129.

C-34. A Purple Star (*Pisaster ochraceus*) in the waters of the Gulf Islands, southern Strait of Georgia. Page 135.

C-35. An aggregation of Purple Stars (*Pisaster ochraceus*) in Barkley Sound. Page 135.

C-36. A Giant Pink Star (*Pisaster brevispinus*) near Sooke, on the southern tip of Vancouver Island. Page 132.

C-37. A Sunflower Star (*Pycnopodia helianthoides*) in Saanich Inlet, just northwest of Victoria, B.C. Page 139.

C-38. The common black form of a Long-rayed Star (*Stylasterias forreri*) among Strawberry Anemones (*Corynactis californica*) off Helby Island, Barkley Sound. Page 143.

C-39. The brown form of a Long-rayed Star (*Stylasterias forreri*) in Tasu Sound, Queen Charlotte Islands. Page 143.

Family Pterasteridae

Characteristics: The aboral side of body is inflated and the oral side flat. A supradorsal membrane supported by the spines of the paxillae (figure 80) covers the true aboral surface to create a nidamental chamber. In the centre of this membrane is a large opening (osculum). Many smaller spiracles pierce the rest of the membrane. On each side of the ambulacral furrow is a wide actinolateral membrane supported by long spines; between the spines are small holes, each guarded by an operculum, which lead to the nidamental chamber; water enters here and is expelled through the osculum. No oral intermediate plates.

In this region:
Diplopteraster multipes
Pteraster militaris
Pteraster tesselatus.

Deeper than 200 metres:
Hymenaster perissonotus
Hymenaster quadrispinosus
Pteraster jordani
Pteraster marsippus.

Hymenaster koehleri is normally found north of this region, but it has also been collected off the coast of California (Chris Mah personal communication), so we may expect it to be found in this region eventually.

Diplopteraster multipes

multipes: **Latin *multus*, much.**

Description

Diplopteraster multipes is a plump, pentagonal sea star with the aboral surface bristling with spines that protrude through the supradorsal membrane. The alternating adambulacral spine arrangement is characteristic of this species. The body is purple-red aborally and greyish-white orally, with reddish-orange tube feet. *D. multipes* has five arms (rarely six) up to 10 cm long. The ratio of arm to disc ranges from 1.2 to 1.6. The aboral supradorsal membrane is pierced by the central spine of each paxilla and 8 to 20 pores or spiracles (figure 80). The central spine is surrounded by 7 to 9 shorter, slender spines. A large pore, the osculum, in centre of aboral surface opens and closes as water is expelled. The number of spines in the transverse combs of the adambulacrals alternates from 3 or 4 per plate to 4 or 5. Every other comb is set back slightly from the furrow (figure 81). The mouth plates have 4 or 5 marginal spines and 1 slender suboral spine.

Taxonomic Note: Originally described as *Pteraster multipes* Sars; revised to *Diplopteraster multipes* by Verrill (1880).

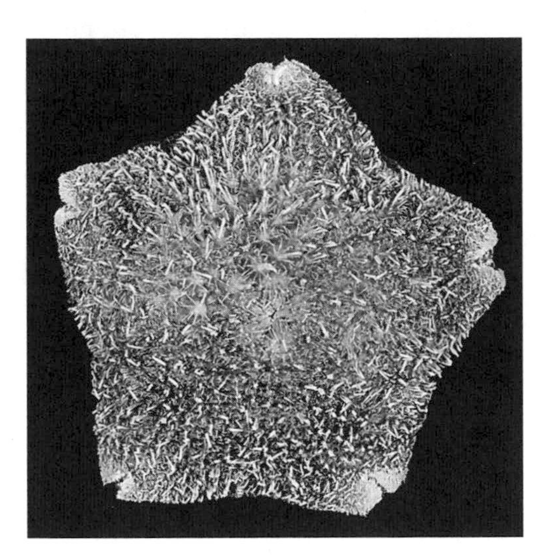

79. *Diplopteraster multipes*, R = 50 mm.

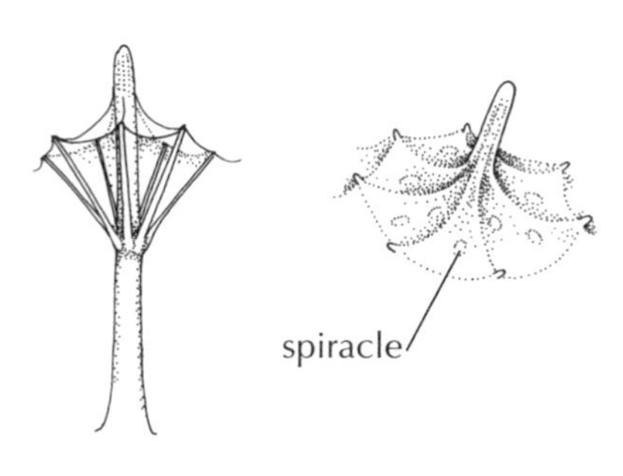

80. Aboral pseudopaxilla supporting the supradorsal membrane.

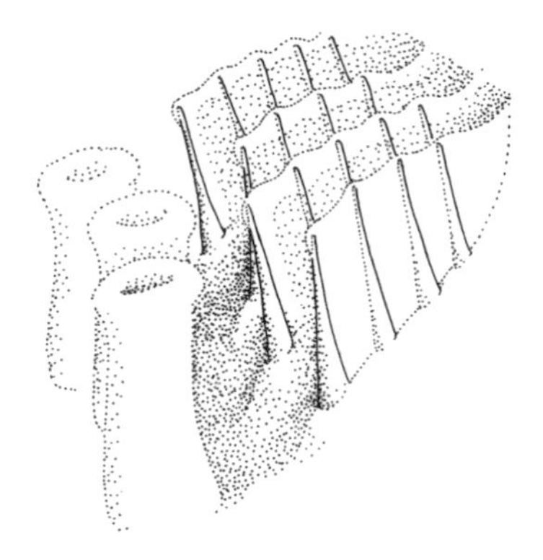

81. Adambulacral spines of *D. multipes.*

Similar Species

Diplopteraster multipes is similar in shape to *Pteraster* but the protruding spines distinguish this species. *Pteraster militaris* and *P. tesselatus* are commonly seen in shallow water, but *D. multipes* is not.

Distribution

Circumpolar; in the Atlantic, from the Arctic to Chesapeake Bay and southwest of Ireland, 91 to 1225 metres deep; in the Pacific, to northern Japan and San Diego, California, at depths of 57 to 1171 metres; and off the coast of South Africa. Found on mud, sand and gravel. Uncommon in this region.

Biology

The stomach contents of collected *D. multipes* included brittle star fragments and sand. Off Oregon, two specimens had stomachs partially everted when collected, suggesting extra-oral digestion of prey, like most sea stars. *D. multipes* secretes copious amounts of phosphorescent mucus.

References

Carey 1972, Harvey et al. 1988. **Range:** Alton 1966, Clark and Downey 1992.

Pteraster militaris **Wrinkled Star**

militaris: **Latin, of soldiers and war.**

Description

Pteraster militaris is a stubby-armed, pale yellow sea star with a wrinkled aboral surface bearing a large pore (osculum) in the centre. Its colour can vary from creamy white to yellow or pink. Its five arms can grow up to 7.5 cm long. The ratio of arm to disc is 2.0 to 2.5. The aboral surface is a soft, fleshy, wrinkled, supradorsal membrane covering the aboral pseudopaxillae, which consist of a low base, slightly higher than it is wide, with 3 to 5 slender spinelets radiating from the top. Under magnification, tiny calcareous bodies can be seen in the membrane but no muscle bands as in *Diplopteraster multipes*. The adambulacrals of *P. militaris* have a transverse webbed comb of 6 to 9 spines, the proximal spine being the smallest. Extending horizontally from the outer end of the plate is a large spine (about three times length of the plate) embedded in the actinolateral membrane (figure 83). The free outer edge of this membrane almost delineates the edge of the arm when viewed from the aboral side. Each mouth plate has 6 to 8 slender marginal spines joined by a membrane. A single large glassy-tipped suboral spine stands near the suture on each plate.

Taxonomic Note: Originally described as *Asterias militaris* O.F. Müller; revised to *Pteraster militaris* by Müller and Troschel (1842).

82. *Pteraster militaris,* R = 44 mm. (See colour photograph C-20.)

Similar Species

Pteraster militaris has longer arms and a softer, more wrinkled skin than its close relative, *Pteraster tesselatus*. The fans of marginal spines on adjoining mouth plates are not joined together by a continuous membrane as in *P. tesselatus*.

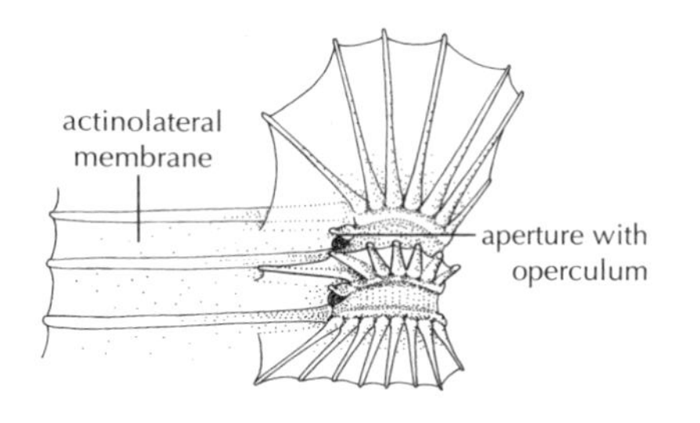

83. Adambulacral spines of *P. militaris.*

Distribution

Circumboreal; in the North Pacific, from the Bering Sea to northern Oregon and to the Sea of Japan; in the North Atlantic, on the Norwegian coast and to Cape Cod, U.S.A. Found at depths of 10 to 1100 metres, usually on mud, but in less than 30 metres on rocky substrates. In the Arctic, it is found on rock and a mixture of rock and mud.

Biology

Pteraster militaris eats the sponge *Iophon pattersoni* and the hydrocorals *Allopora verrilli* and *A. petrograpta* (N. McDaniel personal comment). The Atlantic form begins breeding in October. The female retains some eggs in the nidamental chamber and releases others. Large females brood a smaller proportion of spawned eggs than small females. Sperm enters through ambulacral pores to fertilize the eggs in the chamber. Thirty or forty juveniles grow to an arm's length of 4.5 mm in this chamber. While in the chamber, they feed on maternal skin, faecal material, mucus, aborted larvae and, possibly, dissolved organic matter. The sexes are separate and females have a continuous breeding season with a slight decrease in spring. Male gonads usually ripen in synchrony with those in other males, but female gonads do not.

References

Dembetskii 1988, Falk-Peterson and Sargent 1982, Kaufman 1968, McClary and Mladenov 1988, 1989 and 1990, Yayli 1994.
Range: Alton 1966.

Pteraster tesselatus **Cushion Star, Slime Star**

tesselatus: **Latin, regularly chequered, resembling a mosaic.**

Description

Pteraster tesselatus is a stubby five-armed sea star with a central aboral pore. The ends of the arms are upturned to reveal the ambulacral furrow. When collected this species often produces copious amounts of clear mucus. Usually tan, but its colour varies from cream to yellow to a dull grey-purple. The aboral side is occasionally modified by a chequered pattern or dark brown markings in a rough star-shaped pattern. The tube feet are yellow or orange. The arms can grow up to 12 cm long, and the arm-to-disc ratio is from 1.1 to 1.9. The aboral surface is a supradorsal membrane supported by the pseudopaxillar spines with an opening, the osculum, in the centre. The pseudopaxillae have a low base and up to 26 long spines, and 7 or 8 on the periphery. Marginals and other skeletal plates are hidden by the supradorsal membrane. The adambulacrals have a webbed transverse series of 5 to 7 spines. An actinolateral membrane covers half the distance from the furrow to the edge of the arm. Each mouth plate has a membranous fan of 5 to 7 marginal spines and a single large suboral spine. The fan of one plate is continuous with the fan of the adjacent one (figure 85).

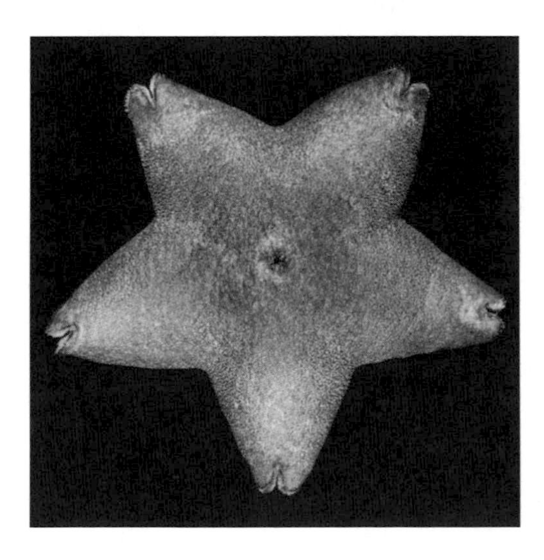

84. *Pteraster tesselatus*, R = 40 mm. (See colour photograph C-21.)

Similar Species

Pteraster tesselatus has shorter arms, a smoother aboral surface and generally is a darker colour than *Pteraster militaris*.

Distribution

The Bering Sea to Washington (subspecies *P. tesselatus arcuatus* to central California) in 6 to 436 metres. Usually found on broken or solid rock along the coast of British Columbia. Common, but not abundant.

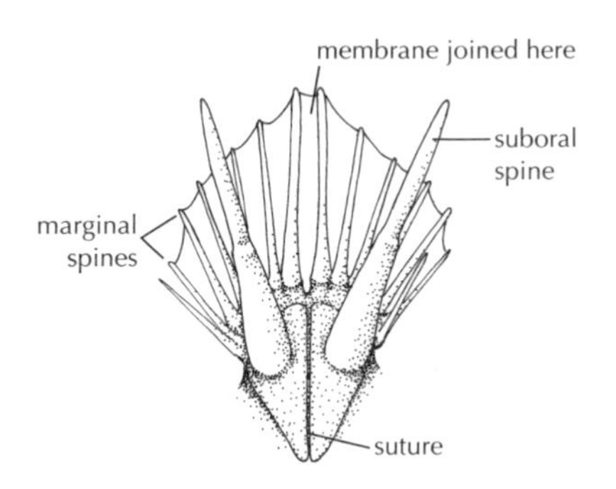

85. Mouth plates of *P. tesselatus.*

Biology

The diet of *Pteraster tesselatus* includes encrusting sponges, bryozoans, hydroids and colonial sea squirts, and less frequently, sea anemones, the rock oyster *Pododesmus macrochisma,* scallops and solitary sea squirts.

P. tesselatus breeds from early July to the end of August. It does not brood its eggs in the so-called nidamental chamber. Instead, it releases yellow or orange eggs (1.5 mm in diameter) through the osculum two or three at a time or in a long string. The fertilized egg develops into a nonfeeding larva that passes directly into the adult without metamorphosis, which usually results in the axes of symmetry changing or the loss of larval structures. This is the only sea star known to have truly direct development. McEdward (1995) suspects that the larval form evolved in a brooding ancestor but did not revert back to a typical pelagic larva when the species evolved back to broadcast spawning. At 10°C in the laboratory, development from egg to young sea star took 25 days; the mouth formed on the 30th day. This implies that, in nature, this sea star does not feed until it settles on the substrate as a juvenile. Most settlement occurs between 10 and 12 days, often among the tubes of the polychaete *Phyllochaetopterus*. See McEdward 1992 for details of the unusual development of this species.

P. tesselatus shows a negative response to bright light and a positive response to moderate or green light. When it is disturbed, cells in its supradorsal membrane secrete copious amounts of mucus, which is forced out of the spiracles onto the top surface by water pressure in the nidamental chamber. This secretion is highly effective in repelling predation by *Solaster dawsoni* and *Pycnopodia helianthoides*; experimental removal of the supradorsal membrane allowed an *S. dawsoni* to eat a *P. tesselatus* in four days. If a clam is injected with *P. tesselatus* mucus and offered to the Sunflower Star, it is rejected. In an aquarium, this mucus will kill snails, hermit crabs and sea cucumbers that are submersed in it for 24 hours.

P. tesselatus flushes its gills by inflating its supradorsal membrane to draw in sea water through slits between the adambulacral spines, and contracting the membrane to force water out through the osculum. This allows the animal to regulate its oxygen intake. The scale worm *Arctonöe vittata* is commensal on *P. tesselatus*.

References

Birkeland et al. 1971, Chia 1966b, Janies and McEdward 1993 and 1994, Johansen and Petersen 1971, Mauzey et al. 1968, McEdward 1992 and 1995, McEdward and Chia 1991, McEdward and Coulter 1987, Nance and Braithwaite 1979 and 1981, Rodenhouse and Guberlet 1946, Stickle 1985c, Tezon et al. 1984. **Range:** Fisher 1911.

Family Korethrasteridae

No species in this region.

Deeper than than 200 metres:
Peribolaster biserialis

ORDER SPINULOSIDA

Family Echinasteridae

Characteristics: The disc is small in relation to the arms, which are long, narrow and cylindrical. The aboral plates are arranged as a fine or coarse mesh, usually bearing short spines, alone or in groups. No pedicellariae. The tube feet have single ampullae.

In this region:
Henricia aspera aspera
Henricia asthenactis
Henricia leviuscula annectens
Henricia leviuscula leviuscula
Henricia leviuscula spiculifera
Henricia longispina longispina
Henricia sanguinolenta.

Henricia

The genus *Henricia* consists of a number of exceedingly variable species. Fisher (1911) describes eight species, five subspecies and a multitude of varieties for the North Pacific and adjacent waters, but admits that his attempts should be regarded as provisional. He states: "Although well acquainted with the variability of starfishes, I have never before met with such an extreme example as the species of this genus present." The genus *Henricia* would make a good subject for a phylogenetic study combining traditional morphological analysis with modern genetic techniques.

I have tried to simplify the species accounts for this book by describing a typical individual of each species, rather than the complete variation. Consult Fisher 1911 for detailed descriptions of all the subspecies and varieties of *Henricia*. Clark (1996) formally designated some of Fisher's species as subspecies; for example, what Fisher called *Henricia leviuscula* Clark revised to *Henricia leviuscula leviuscula* to differentiate it from *H.l. annectens.*

Henricia aspera aspera **Ridged Blood Star**

aspera: **Latin *asper*, rough, harsh.**

Description

Henricia aspera aspera is a long-armed sea star with a coarse, meshwork of ridges on the aboral surface bearing groups of 5 to 15 small, sharp spinelets, and with large papular areas between the ridges (figure 87). The body colour is yellow to brick red with a paler oral side. The five thin arms are up to 16 cm long, and the arm-to-disc ratio is from 5.3 to 7.2. The supero- and inferomarginals form an obvious double series along the arm. The intermarginals form a triangular patch in the proximal one-fifth of the arm. The oral intermediates occur proximally for three-quarters of the arm. The adambulacrals have 1 or 2 spinelets deep in the furrow with 2 to 8 coarse spinelets in a transverse zigzag or double row on the oral surface, the two nearest the furrow being the largest (figure 88).

Taxonomic Note: Originally decribed as *Henricia aspera* Fisher; revised to *Henricia aspera aspera* by Clark (1996).

86. *Henricia aspera aspera*, R = 115 mm. (See colour photograph C-22.)

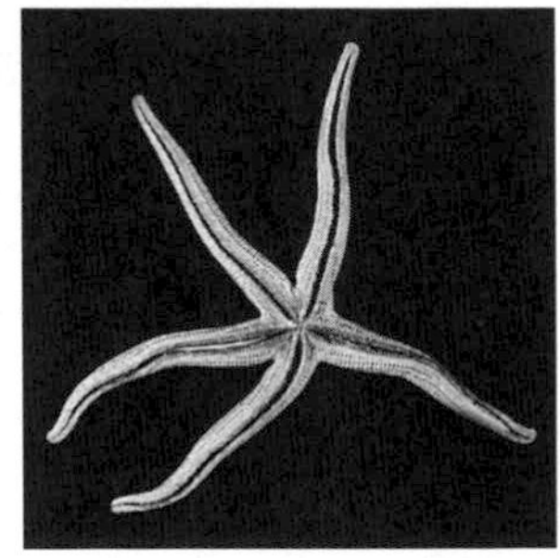

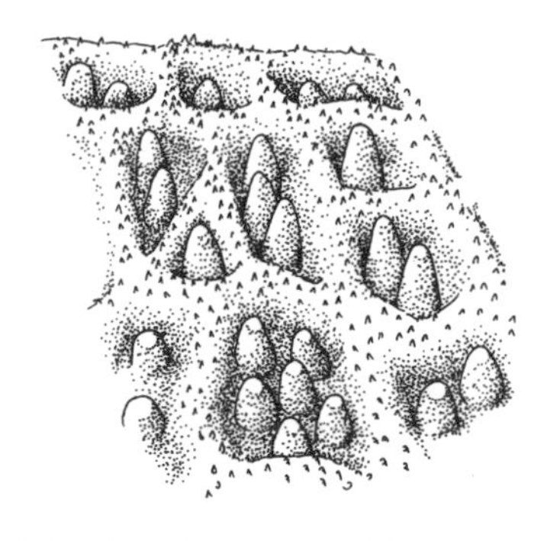

87. Aboral surface of *H. aspera.*

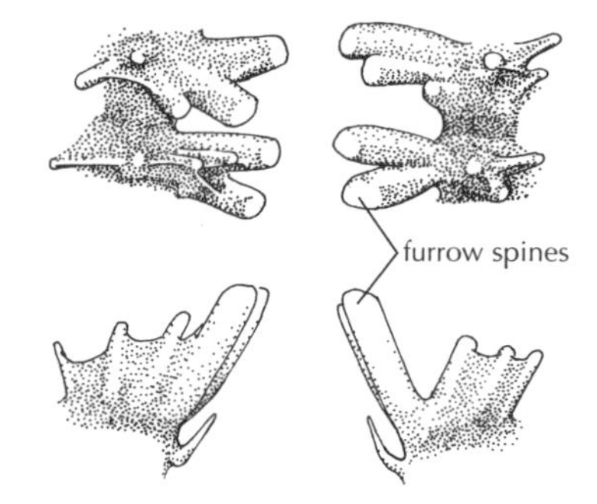

88. Adambulacral spines.

Similar Species

Henricia aspera aspera can be distinguished from other *Henricia* species by its large size, the open meshwork of ridges on the aboral surface bearing groups of small spinelets, and the small number of adambulacral spines.

Distribution

The Bering Sea to Santa Barbara, California, and to the Sea of Japan. Found on mud, sand, pebbles and rock at depths of 6 to 904 metres. Uncommon in this region. I have collected it on rock while scuba diving in Desolation Sound and Barkley Sound, and off Lasqueti Island and the Queen Charlotte Islands.

Biology

Nothing is known about the biology of this species.

References

McDaniel et al. 1978. **Range:** Lambert 1978b.

Henricia asthenactis

asthenactis: **Greek *astheno*, weak, referring to the open, meshlike aboral skeleton.**

Description

Henricia asthenactis has a meshlike aboral surface with a single row of 2 to 4 flesh-covered, well-spaced spinelets on the ridges of each plate (figure 90). The areas between the ridges have about 10 papulae. In alcohol, specimens are yellowish-white. The five arms are up to 11 cm long, and the arm-to-disc ratio is from 4.4 to 6.6. There is a regular series of supero- and inferomarginals with spinelets on each plate and a few intermarginals with 1 or 2 spinelets, at the base of each arm. The oral intermediates form a short series proximally. The adambulacrals have a transverse series of 3 to 5 spines, the proximal spine being longer than the width of the plate, and there is a small knoblike spinelet deep inside the furrow (figure 91).

Similar Species

The aboral surface of *Henricia asthenactis* resembles *H. aspera aspera*, but the arms are shorter relative to the disc and the spinelets on the ridges are larger.

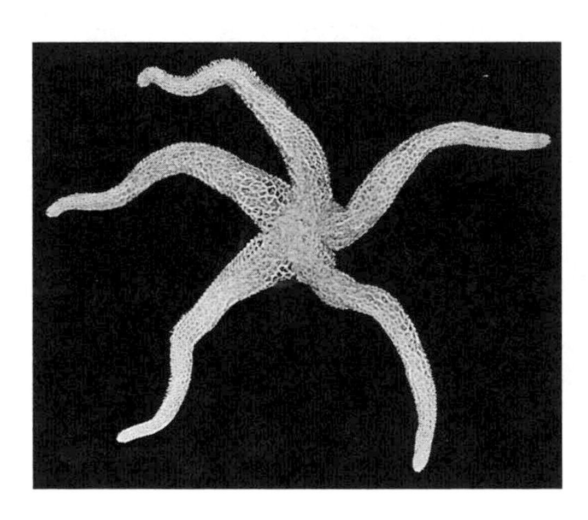

89. *Henricia asthenactis*, R = 53 mm.

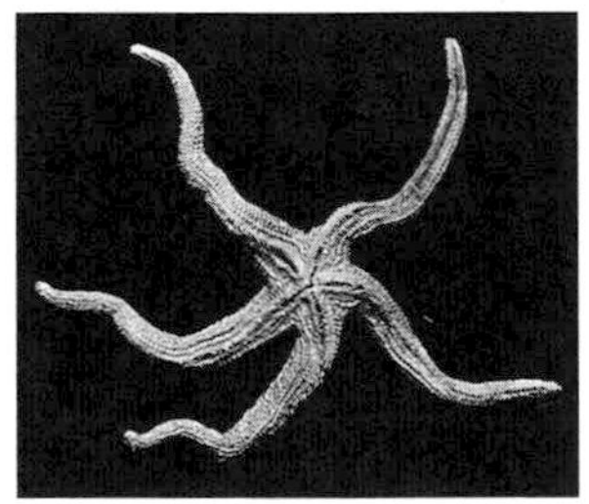

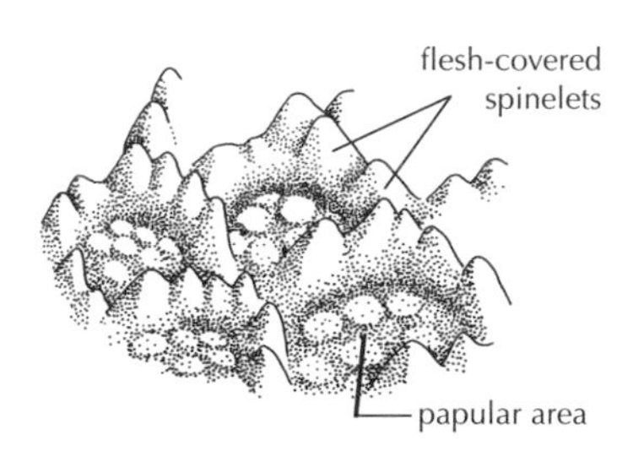

90. Aboral surface of
H. asthenactis.

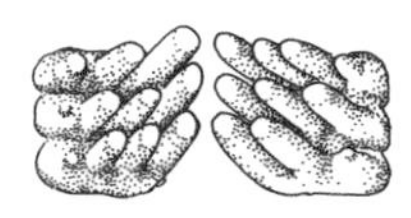

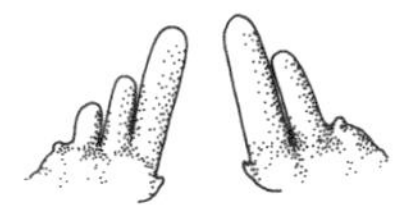

91. Adambulacral spines
(top view above and
side view below).

Distribution

Kamchatka, the Shumagin Islands, the Bering Sea to Santa Barbara Island, California and Gulf of California. Found on mud, sand or gravel at depths of 91 to 1250 metres. Rare in this region.

Biology

Nothing is known about the biology of this species.

References

Fisher 1911, Lambert 1978b. **Range:** Clark 1996.

Henricia leviuscula leviuscula **Blood Star**

Other subspecies:

H. leviuscula annectens
H. leviuscula spiculifera = *H. leviuscula multispina*

leviuscula:	**Latin *levis*, smooth.**
annectens:	**Latin, linking, joining.**
multispina:	**Latin *multus*, much, and *spina*, thorn.**
spiculifera:	**Latin *spiculum*, a sharp point.**

Description

Henricia leviuscula leviuscula is the most common red thin-armed sea star in the intertidal and shallow waters of British Columbia. The colour may vary from orange to brick-red, to brown, and some specimens may have a grey patch around the base of the arms or covering the disc. The aboral surface is variable, but the three rows of plates along the lower side of each arm are a constant feature. *H.l. leviuscula* has five arms (rarely six) up to 16 cm long, and an arm-to-disc ratio of 3 to 7. The aboral plates, covered with 30 to 60 small, close-set spinelets, are close together and normally larger than the papular spaces. The supero- and inferomarginals form a regular series. At the base of each arm, the two marginal series split to form a triangular group of intermarginals. The oral intermediates form a regular series to the arm tip, or nearly so. The adambulacrals bear 5 to 12 stubby spinelets in two transverse rows, increasing in size proximally, and a short, curved spinelet deep in the furrow (figure 92).

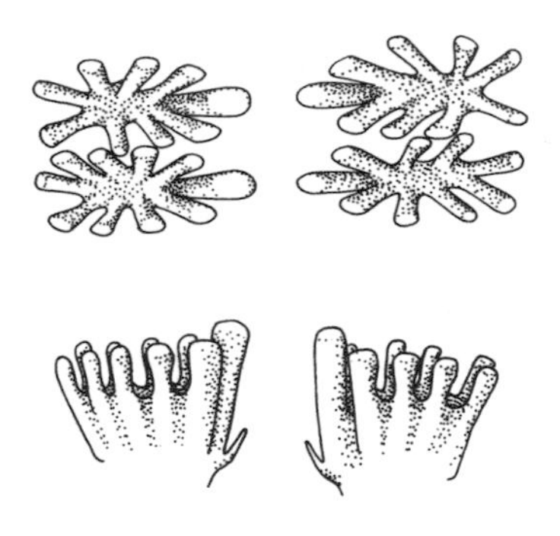

92. Adambulacral spines of *H. leviuscula*.

The subspecies, *H.l. annectens*, is intermediate in appearance between *H. aspera aspera* and *H.l. leviuscula*. The aboral pseudopaxillae have 5 to 20 stubby spinelets and the oral intermediates form a series one-half to two-thirds of the arm length.

The pseudopaxillae of *H.l. multispina* are more densely

packed than those of the common intertidal *H.l. leviuscula*, with 40 to 60 bristling glassy spinelets. The intermarginals vary from a triangular area to one-half of the arm length. The adambulacrals have 25 to 40 multidenticulate spinelets.

Small, yellowish sea stars that live under rocks in the intertidal zone and brood their eggs are probably an undescribed species of *Henricia* (see colour photograph C-25). Meg Strathmann of Friday Harbor Laboratories is working on the description and naming of this new animal.

Taxonomic Note: Originally described as *Linckia leviuscula* Stimpson; revised to *Henricia leviuscula* by Fisher (1910) and *Henricia leviuscula leviuscula* by Clark (1996).

Distribution

Henricia leviuscula leviuscula: the Aleutian Islands to Turtle Bay, Baja California; intertidal to 400 metres. The typical form of *H.l. leviuscula* is common on rocky substrates in shallow water. *Henricia leviuscula annectens*: the Gulf of Alaska to Santa Barbara, California; 10 to 228 metres. *Henricia leviuscula spiculifera*: the Bering Sea to Oregon and to Okhotsk and the Sea of Japan; 9 to 680 metres.

93. *Henricia leviuscula leviuscula*, R = 73 mm. (See colour photographs C-0 and C-23.)

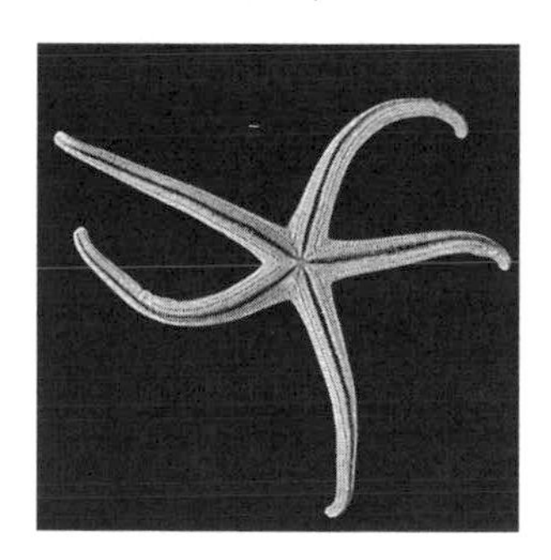

94. A male Blood Star spawning in late April. The sperm it has released appears as a milky cloud curling between two arms.

Biology

Henricia leviuscula leviuscula feeds on plankton, dissolved organic matter, bryozoans and the encrusting sponges *Sigmadocia* sp. and *Isodictya quatsinoensis* (N. McDaniel personal comment).

I have seen large males releasing sperm from gonopores between the arms in April and May (figure 94). Small yellowish females with brown mottling, that brood bright red-orange eggs (500 micrometres in diameter) from January through March are probably an undescribed species of *Henricia* (M. Strathmann, personal communication). The scale worm *Arctonöe vittata* and the caprellid *Caprella greenleyi* are commensal on *H.l. leviuscula*. Juveniles, less than 10 mm, live on tubes of the polychaete worm *Phyllochaetopterus*. *H.l. leviuscula* rights itself by somersaulting: it raises its central disc as it presses its arms onto the substrate, then two of the arms "walk" underneath the disc and the opposing arms fold over, so that the animal lands on it oral surface.

References

Birkeland et al. 1971, Eakin and Brandenburger 1979, Ferguson 1967 and 1994, Hopkins and Crozier 1966, Lissner and Hart 1996, Manchenko 1987, Mauzey et al. 1968, McCain 1969, Polls and Gonor 1975, Simoncini and Moody 1990, Strathmann 1987, Van Veldhuizen and Oakes 1981. **Range:** Clark 1996, Fisher 1911, Lambert 1978a, Maluf 1988.

Henricia longispina longispina

longispina: **Latin *longus*, long, and *spina*, thorn.**

Description

Henricia longispina longispina has a coarse meshlike skeleton that bears, at regular intervals, radiating clusters of 2 to 9 long, sharp spinelets. These spinelets are longer (1 to 1.5 mm) and stouter than any other species of *Henricia*. The body is milky white. It has five arms up to 10 cm long, and an arm-to-disc ratio of 3.1 to 5.2. Papulae occur in groups of five or six. A regular series of supero- and inferomarginals reach to the tip of the arm and each has 6 to 10 spinelets. The intermarginals are smaller and run from one-half to two-thirds the length of the arm. Small oral intermediates run halfway along the arm and each plate has 2 to 4 spinelets. The adambulacrals have a zigzag transverse series of 6 or 7 long, bristly spinelets, decreasing in size distally (figure 95).

95. Adambulacral spines of *H. longispina* (top view above and side view below).

Taxonomic Note: Originally described as *Henricia longispina* Fisher; revised to *Henricia longispina longispina* by Clark (1996).

96. *Henricia longispina longispina*, R = 40 mm.

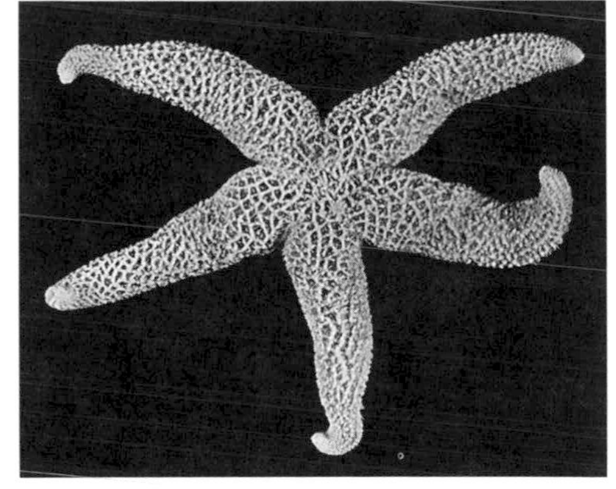

Similar Species

The coarse meshlike skeleton of *Henricia longispina longispina* is similar to *H. asthenactis* and *H. aspera aspera*, but the details of the spinelets and body proportions should separate them.

Distribution

The Bering Sea to the Sea of Okhotsk and Vancouver Island. Found on pebbles or mud-sand at depths of 28 to 512 metres. Uncommon in this region.

References

D'Yakonov 1950, Fisher (1911). **Range:** Clark 1996.

Henricia sanguinolenta **Fat Henricia**

***sanguinolenta*: Latin *sanguis*, blood, and *lentus*, full of, prone to.**

Description

On the Pacific coast, *Henricia sanguinolenta* is nearly always white to pale orange. The Atlantic form is red, as the species name suggests. It has five arms up to 13 cm long, and an arm-to-disc ratio of 5 to 7. The aboral surface consists of small, close-set pseudopaxillae with few (3 to 6) to many (25 to 30) spinelets. The papular areas are small, with 1 to 4 papulae. The marginals are also small and do not form an obvious series. A row of intermarginals varies from a short series to the full length of the arm. Oral intermediates form one or two series, with the longest row reaching to the distal third of the arm. The adambulacrals usually have 10 to 15 (5 to 25) spinelets on each plate, 2 or 3 of these on the edge of the furrow. A short curved spinelet sits on the furrow face (figure 98).

Taxonomic Note: The taxonomy of this species in the North Pacific is in need of revision. There is disagreement in the literature that remains unresolved. According to Madsen (1987), *H. sanguinolenta* does not occur in the North Pacific, but its close relative, *H. tumida* Verrill, 1909, does. *H. tumida* has short stubby arms and its arm-to-disc ratio is from 2 to 2.7. Fisher (1911) examined over 1000 specimens of *Henricia* from the west coast and determined that

97. *Henricia sanguinolenta*, R = 100 mm. (See colour photograph C-24.)

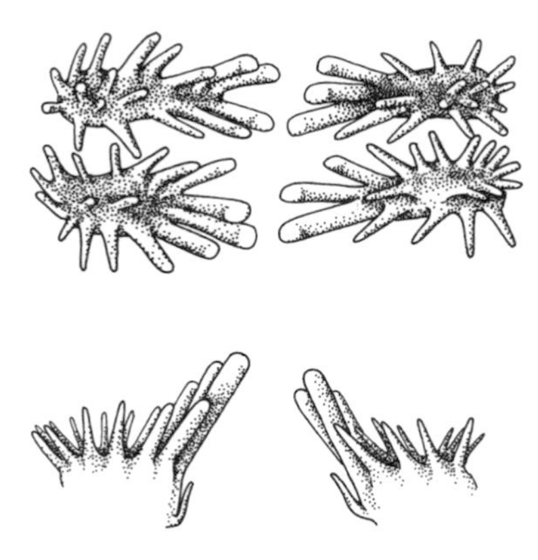

98. Adambulacral spines of *H. sanguinolenta* (top view above and side view below).

many fit the description of *H. sanguinolenta*. I have also examined specimens with an arm-to-disc ratio of 4.5 to 6.9, which is well out of the range of *H. tumida* and closer to that of *H. sanguinolenta*. Madsen did not consider North Pacific specimens in his study, so I am more inclined to agree with Fisher for the purposes of this book. Obviously, more taxonomic work on this species is needed.

Similar Species

In contrast to other *Henricia* species, this one has small marginals that do not form an obvious series. The arms are generally fat at the base and often there is a crease where the arm joins the disc. Within diving depth this is the only white *Henricia*.

Distribution

Circumpolar. To Cape Hatteras in the Atlantic and Washington in the Pacific. Found on solid rock or mud at depths of 15 to 518 metres in the Pacific and 200 metres in the Atlantic. I have often seen it draped over boot sponges in fiords. More common in mainland inlets of British Columbia.

Biology

The biology of the Pacific form of *Henricia sanguinolenta* has not been studied. In the Atlantic form, a complex series of grooves, called Tiedemann's diverticula, connect the stomach to five pairs of digestive glands. The cilia lining these grooves create a current that allows the sea star to feed on plankton or suspended particles. *H. sanguinolenta* is frequently found on the sponges *Mycale, Ficulina* and *Hymeniacidon* and was thought to eat them, but Anderson (1960) suspected that they were tapping into the feeding current of the sponge. The sponge seldom seems to be damaged much by the sea star. This habit may explain why we see it here associated with Boot Sponges.

H. sanguinolenta breeds from February to April on the Atlantic coast. In the White Sea, breeding begins at the end of June. Males cast sperm into water, females arch up and retain fertilized eggs around the mouth. Twenty days after postembryonic development begins, the larva transforms into a juvenile sea star. This species shows a negative response to light. The parasitic copepod *Asterocheres lilljeborgi* lives on its skin.

References

Anderson 1960, Burnell et al. 1986, Collier 1983, Ferguson 1967, Ferguson and Walker 1991 and 1993, Ferrand 1982, Holme 1966, Kaufman 1968, Lafay et al. 1995, Madsen 1987, Rasmussen 1965, Sheild and Witman 1993. **Range:** Clark and Downey 1992.

ORDER FORCIPULATIDA

Family Zoroasteridae

No species in this region.

Deeper than than 200 metres:
Myxoderma sacculatum
Zoroaster evermanni
Zoroaster ophiurus

Family Asteriidae

Characteristics: Five or more arms. At least one adambulacral is fused into an adoral carina. The adambulacrals are wider than their length. Crossed and straight pedicellariae are present, the former usually in dense tufts around the spines. The aboral skeleton is meshlike. The tube feet are arranged in four rows.

In this region:
Evasterias troschelii
Leptasterias aequalis species complex
Leptasterias alaskensis species complex
Leptasterias coei
Leptasterias hexactis species complex
Leptasterias polaris katherinae
Lethasterias nanimensis
Orthasterias koehleri
Pisaster brevispinus
Pisaster ochraceus
Pycnopodia helianthoides
Stylasterias forreri

Deeper than 200 metres:
Stephanasterias albula (one record in Lynn Canal).

Evasterias troschelii **Mottled Star**

troschelii: **after the German taxonomist Franz Hermann Troschel, 1810–82.**

Description

Evasterias troschelii has a fairly distinctive colour, usually mottled green, red, brown or orange, with white spines and a lighter oral surface. It has five arms up to 30 cm long, and its arm-to-disc ratio ranges from 5.0 to 7.6. This is one of the most variable sea stars in this region, especially in the pattern of aboral spines, which varies from numerous small spines in a netlike pattern (figure 99) to longitudinal rows of large blunt spines on the arms (figure 100). Fisher (1930) described three forms to cover this range of variability, but they are not geographically separated so are not considered subspecies. Papulae occur abundantly between the spines. A regular series of superomarginals curves upward at the junctions of the arms and meets the series from the adjoining arm. The first two rows of similar spines below the superomarginals are inferomarginals, the next four rows are oral intermediates (figure 101). The adambulacrals alternately bear 1 and 2 spines. Three to five pairs of adambulacrals from adjoining arms fuse together to form an adoral carina. The mouth plates at the proximal end of the adoral carina bear two unequal marginal spines and one longer suboral spine. Lanceolate pedicellariae occur on the oral and adambulacral spines (figure 101). Crossed pedicellariae occur on the aboral and marginal spines and also in clusters on the oral and adambulacral spines.

99. *Evasterias troschelii* (small-spined form), R = 210 mm. (See colour photograph C-26.)

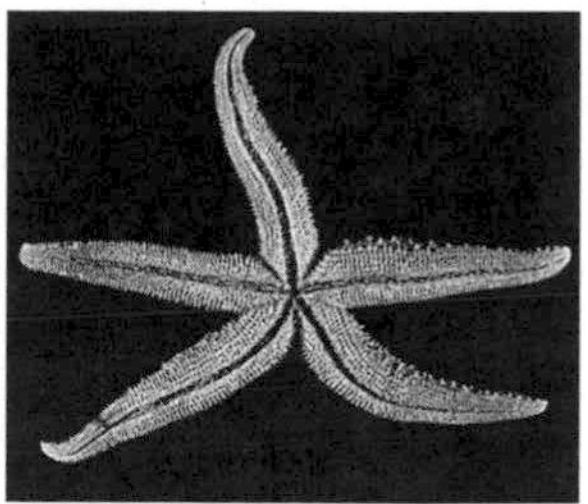

100. *Evasterias troschelii* (large-spined form), R = 190 mm. (See colour photograph C-26.)

Taxonomic Note: Originally described as *Asterias troschelii* Stimpson; revised to *Evasterias troschelii* by Verrill (1914).

Similar Species

Evasterias troschelii is similar to *Pisaster ochraceus*, but it has a greater arm-to-disc ratio, more rows of spines between the marginals and ambulacral furrow, and a less sunken mouth. The six-armed star, *Leptasterias hexactis*, although smaller, might be confused with a juvenile *E. troschelii*.

Distribution

The east coast of Kamchatka and the Pribilof Islands to Monterey, California. Usually found on rock or cobble, but also on sand, from the intertidal zone to 75 metres deep. Common on the northwest coast of North America, especially in the more sheltered waters of inlets. *Evasterias troschelii* tends to replace *Pisaster ochraceus* in these localities.

Biology

Evasterias troschelii has a varied diet, including bivalves (especially the rock oyster *Pododesmus machrochisma* around the San Juan Islands), barnacles (in Gabriola Passage), sea squirts, chitons, gastropods and lamp shells. On rocky shores, *E. troschelii* eats mostly molluscs and barnacles, while rejecting tunicates; but on soft bottoms, it eats tunicates with a preference for those with thinner tunics. It digests tough-skinned tunicates, like *Styela gibbsii* and

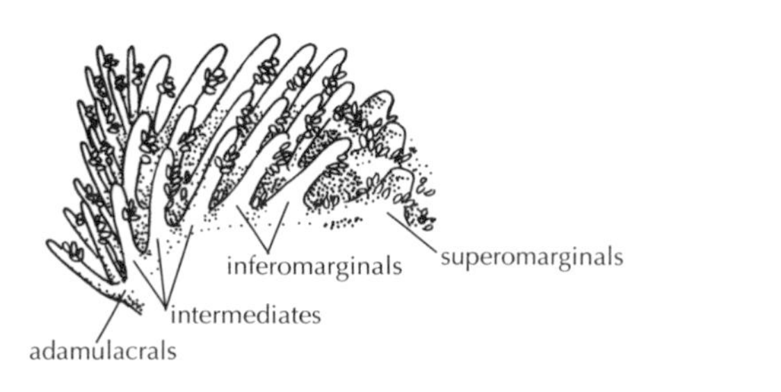

101. Oral arm spines (left) and adambulacral spines (right) of *E. troschelii.*

Pyura haustor, through a hole in the tunic. To eat a bivalve, *E. troschelii* everts its stomach and inserts it between the shells while the tube feet pull from the outside. A specimen with arm length of 22.5 cm can exert an average force of 4500 grams for six hours and a maximum force of 5500 grams. A small concentration (0.12 parts per million) of the water soluble fraction of crude oil reduces the feeding rate of *E. troschelii*; above 0.97 ppm, feeding ceases. In Southeast Alaska, an Alaska King Crab (*Paralithodes camtschatica*) was reported to attack and eat *E. troschelii.*

Around Vancouver Island and in Southeast Alaska this species breeds from April to June. Eggs are numerous and small (150 micrometres). Gastrulation takes place on the second or third day after fertilization. By the fourth day a bipinnaria larva forms.

A parasitic ciliate *Orchitophrya stellarum*, endemic to the North Atlantic, was discovered recently in *E. troschelii*. It was first discovered in 1988 in the Purple Sea Star (*Pisaster ochraceus*) in British Columbia, but appears to have spread to this other member of the Family Asteriidae. It infests the male gonad causing castration and loss of sperm production. The scale worm *Arctonöe fragilis* is commensal on *E. troschelii.*

References

Byrne et al. 1998, Christensen 1957, Fisher 1930, Mauzey et al. 1968, Meijer et al. 1987, Meijer and Zarutskie 1987, Mortensen 1921, O'Clair and Rice 1985, Patterson et al. 1978, Pelech et al. 1987, Pelech et al. 1988, Sloan and Robinson 1983, Smith 1982, Smith et al. 1982, Stickle 1985c, Stickle et al. 1992, Strathmann 1971, Turner et al. 1987, Young 1985. **Range:** D'Yakonov 1950, Fisher 1930, Lambert 1999.

The species of *Leptasterias* are highly variable and difficult to separate on the basis of morphology. Fisher (1930) described 18 species and subspecies of *Leptasterias* from the North Pacific and adjacent waters but, as a caution, he said: "Among the numerous small six-rayed sea stars of the northwest coast of North America, specific lines are exceedingly difficult to draw; but nowhere is there such a confusion and intermingling of forms as in the sounds of Washington and British Columbia."

According to Fisher, six of those species and subspecies as well as seven formae occur in the area covered by this book. He also revised many of the names used earlier by Verrill (1914) and made them synonymous. An identification key to this genus by D'Yakonov (1950) covers many species and subspecies of the Arctic Pacific but not the west coast of North America. Later, Chia (1966c), attempted to clarify the taxonomy of this genus on the west coast using embryology, ecology and behaviour. He concluded that *L. hexactis* and *L. aequalis* in Puget Sound were synonymous, with the former taking precedence. In the first edition of this book (Lambert 1981), I followed his conclusion and used *L. hexactis*. But in the 1980s, Bill Stickle and his collaborators, using electrophoresis and DNA analysis, discovered many genetic lineages (called haplotypes) within the *Leptasterias* of the west coast. This work suggested the possibility of numerous species, but it was unclear if these lineages were equivalent to the various taxonomic species and subspecies that Fisher described in 1930.

Recently J.M. Flowers with his supervisor, D.W. Foltz, have continued this work on the genetics and taxonomy of the intertidal six-armed *Leptasterias* (subgenus *Hexasterias*) along the North American Pacific coast. Their work has helped to clarify the relationship of the 16 known haplotypes, to Fisher's original type descriptions. They found no simple one-to-one relationship of haplotype to taxonomic grouping. In some cases a single haplotype could be represented by multiple taxonomic forms or, conversely, there were multiple haplotypes within what appeared to be a single morphological taxon. Flowers (1999) concluded that, below the species level, Fisher's subspecies and forms did not clearly correspond to known haplotypes. As a result, Flowers has proposed a set of species complexes within which many of the existing subspecies and forms have been grouped, pending further investigation. I have borrowed freely from Jon Flowers' thesis in organizing this section and I thank him for creating some order out of the chaos. Further work on this genus is needed to clarify the relationships of other described taxa in *Leptasterias*. Please consult the literature by these two authors in the next few years. Perhaps I can report on progress in the next edition of this book.

This section is arranged a little differently than the others in this book. Due to the confusion associated with this genus, it is not clear from the literature which information belongs with which species. Hence, the morphological characteristics of each species complex are presented first, with their known geographic ranges taken from Flowers 1999. Then, information about habitat and biology for all the small intertidal brooding species is combined at the end. I have also included two other described species, *L. coei* and *L. polaris*, not dealt with in Flowers 1999 that are larger and seem fairly distinct, but their true status is yet to be confirmed by genetic analysis.

The common name, Six-armed Star, applies to all the *Leptasterias* in this region.

Leptasterias aequalis species complex

aequalis: **Latin *aequalis*, same, uniform.**

Description

Members of the *Leptasterias aequalis* species complex are small to medium sized. They have six arms up to 5 cm long, and their arm-to-disc ratio is about 4. To the naked eye, the aboral and lateral surfaces seem to be compactly covered with clusters of granular spines (up to 10) with small papular areas separating them. These sea stars have a carinal series of tightly spaced clusters of spines and a dorsolateral series of one to six spinelets per cluster in well-defined longitudinal rows; their pedicellariae are usually present but inconspicuous; the aboral spines are typically subcapitate; the superomarginals each have 4 to 8 slightly bent spines in a vertical series; the inferomarginals have two longer spines; and the oral intermediates are arranged in one, occasionally two, well-developed series. The adambulacrals have one or two spines in irregular alternation. There are clusters of crossed pedicellariae on the distal sides of the adambulacrals, oral intermediates and inframarginals.

Taxonomic Note: This species complex includes *Leptasterias aequalis f. aequalis, L. aequalis f. nana, L. hexactis* (part) and *L. aequalis* (part).

102. *Leptasterias aequalis*, R = 30 mm. (See colour photograph C-27.)

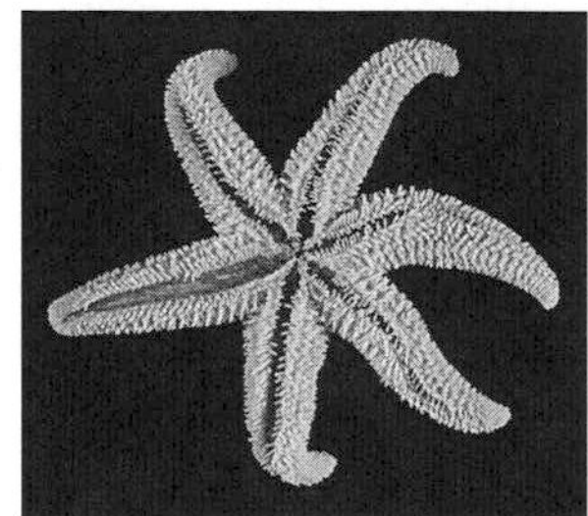

Leptasterias alaskensis species complex

alaskensis: **for Alaska, where many of these species were first described.**

Description

Members of the *Leptasterias alaskensis* species complex have six arms up to 5.5 cm long and an arm-to-disc ratio of 3.3 to 4.5. They have an irregular pattern of robust, capitate aboral spines, less robust in southern Alaska and B.C., but no obvious carinals or longitudinal arrangement of spines. The lateral and oral sides of the arms have broad ovate bivalve pedicellariae. The superomarginals usually have two or three spines, occasionally one and rarely four, on each plate, low on the side of the arm. The inferomarginals are longer and usually in a double row. These species usually have no oral intermediates, but if they are present, they are severely underdeveloped. As with all *Leptasterias,* these have a cluster of crossed pedicellariae on the distal side of the adambulacral spines.

Taxonomic Note: This species complex includes *Leptasterias alaskensis f. alaskensis, L. alaskensis asiatica, L. alaskensis multispina, L. alaskensis f. pribilofensis, L. alaskensis f. shumaginensis, L. hexactis* (part) and *L. hexactis f. aspera.*

103. *Leptasterias alaskensis,* R = 34 mm. (See colour photograph C-28.)

Leptasterias hexactis species complex

hexactis: **Greek *hex*, six, and *aktis*, ray, referring to the six arms of these sea stars.**

Description

Members of the *Leptasterias hexactis* complex have six arms 4 to 8 cm long, each with a distinct carinal row whose plates bear one to three spines apiece. This group of sea stars has arm-to-disc ratios ranging from 3.3 to 5.4. It generally appears to have short stubby arms with an obvious row of light coloured spines down the centre of the arm and widely spaced capitate, dorsolateral spines irregularly arranged in a loose reticulate pattern. The superomarginals have one or two spines, rarely three, and the inferomarginals two. The oral intermediates are usually in a single well-developed series extending beyond half the arm, but never two rows. The adambulacrals have one or two spines, with clusters of crossed pedicellariae on their distal sides (figure 105).

Taxonomic Note: This species complex includes *Leptasterias hexactis f. plena, L. hexactis f. siderea, L. hexactis f. regularis, L. hexactis f. aspera, L. hexactis f. hexactis, L. hexactis vancouveri, L. hexactis* (part) and *L. epichlora*.

104. *Leptasterias hexactis*, R = 27 mm. (See colour photograph C-29.)

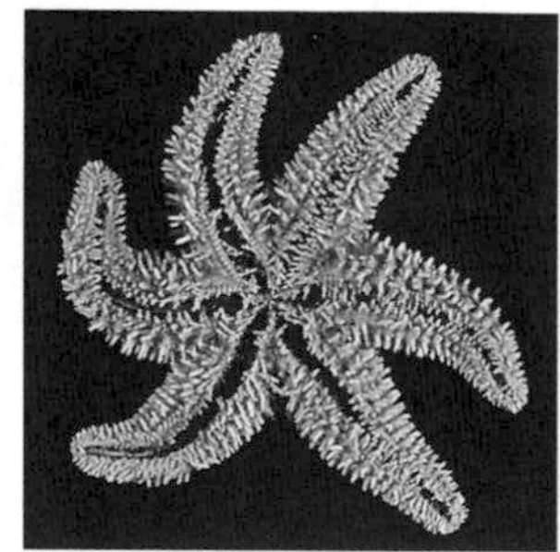

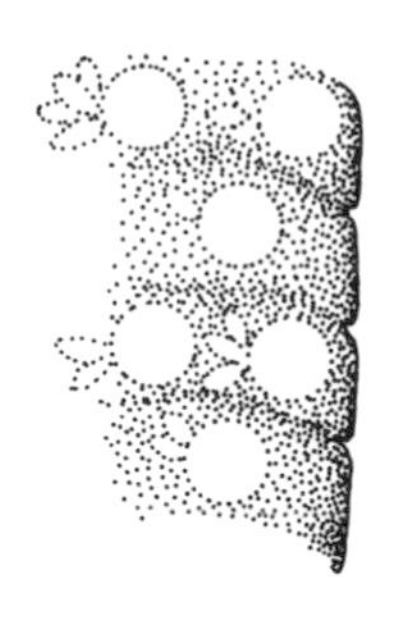

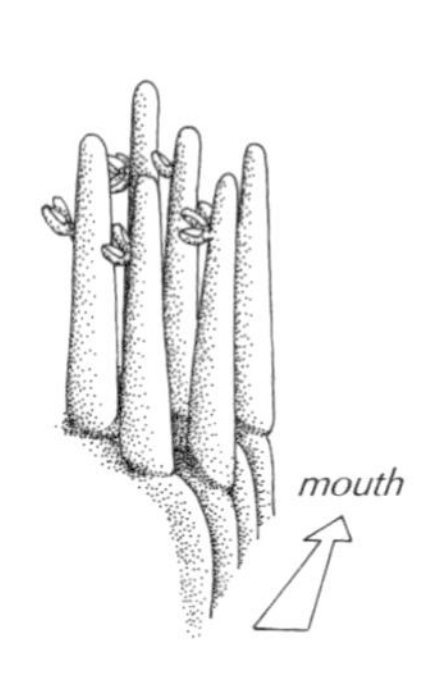

105. Adambulacral spines of *Leptasterias* spp. (top view left and side view right).

Species Similar to *Leptasterias*

Small *Leptasterias* may be confused with juveniles of *Pisaster ochraceus* or *Evasterias troschelii*, which occasionally have six arms. *P. ochraceus* has a single spine on each adambulacral, with a cluster of pedicellariae at the base but not on the spine. *Leptasterias* has one or two spines per plate with a cluster of pedicellariae on the spine itself. *E. troschelii* has pedicellariae on the adambulacrals but an arm-to-disc ratio of 5.0 to 7.6 and six similar rows of spines between the superomarginals and the furrow, made up of two inferomarginals and four oral intermediates. *Leptasterias* has no more than two oral intermediates.

Distribution

Leptasterias aequalis: Vancouver Island and the San Juan Islands to Santa Catalina Island, California, in the intertidal zone. *Leptasterias alaskensis*: the Pribilof Islands to northern Vancouver Island. *Leptasterias hexactis*: The Alaskan Peninsula to the Juan de Fuca Strait, Puget Sound and the Strait of Georgia. The small brooding forms of *Leptasterias* are commonly found on rocky shores under rocks, in crevices and in mussel beds; larger specimens are found subtidally.

Biology

Due to the taxonomic confusion that has existed for many years in the identification of *Leptasterias*, it is not clear if feeding observations in the literature attributed to a particular species are correct. Consequently, I am presenting all the biological observations that refer to intertidal species of *Leptasterias*.

Depending on place, time of year, and abundance of prey, the diet of *L. hexactis* includes barnacles (*Balanus*), sea cucumbers

(*Cucumaria*), limpets (*Acmaea*), snails (*Thais, Littorina* and *Calliostoma*) and chitons (*Katharina, Tonicella*). It does not appear to detect prey from a distance. When prey is abundant, the sea star is more selective for larger individuals. The top snail *Calliostoma ligatum* shows an effective escape response to *L. hexactis*. The commensal slipper limpet *Crepidula adunca* attaches itself almost exclusively to *C. ligatum*, thereby taking advantage of its predator avoidance behaviour. The small pulmonate *Onchidella borealis* has glands along the edge of its body that repel *L. hexactis*. The *Lacuna marmorata* snail reacts to *L. hexactis* by twisting its shell, waving tentacles and dropping off the eelgrass blade that it was attached to. The Carinate Dovesnail (*Alia carinata*) repels *L. hexactis* by nipping its tube feet. The *Balanus glandula* barnacle appears to detect the chemical cues of this predator and stays closed much longer than if mechanically stimulated. *L. hexactis* tolerates low salinity (median tolerance limit: 12.9 parts per thousand) but its feeding and respiration rates drop in this environment.

L. hexactis aggregates to breed in January. Males discharge sperm into the water and females retain fertilized eggs under their arched bodies. The female produces from 50 to 1500 large yellow or orange eggs, which she broods for three months before the juvenile sea stars leave to lead independent lives. The scale worm *Arctonöe fragilis* is commensal on *L. hexactis*.

References

Birkeland et al. 1982, Chia 1966a, 1966c, 1968a and 1968b, Ferguson 1990 and 1994, Fisher 1930, Fishlyn and Phillips 1980, Flowers 1999, Foltz 1997 and 1998, Foltz and Stickle 1994, Foltz et al. 1996a, Foltz et al. 1996b, Francour 1997, George 1994, Highsmith 1985, Hoffman 1980, Hrincevitch and Foltz 1996, Kent 1981, Kwast et al. 1990, Lambert 1994, Lansman 1983a and 1987, Mauzey et al. 1968, Meijer et al. 1987, Menge 1972 and 1975, Menge and Menge 1974, Mladenov et al. 1989a, Moody 1985, Moody and Bosma 1985, Moody and Lansman 1983, Palmer et al. 1982, Pearse and Beauchamp 1986, Polls and Gonor 1975, Shirley and Stickle 1982a, b and c, Stickle 1985a, b and c, Stickle et al. 1982, Stickle et al. 1992, Thomas and Hermans 1985, Van Veldhuizen and Oakes 1981, Vermeij et al. 1987, Webster 1975, Young et al. 1986.
Range: Flowers 1999.

Leptasterias coei

coei: **after W.R. Coe, who collected the original specimens.**

Description

Leptasterias coei is a large sea star with six slender tapered arms up to 14 cm long and an arm-to-disc ratio of 6.5. Alive it is a uniform iron-rust colour. Dried specimens I saw in the Auke Bay collections were beige-orange. The aboral spines are well spaced, coarse, cylindrical and blunt, and surrounded by broad wreaths of crossed pedicellariae. Juveniles have an irregular carinal row, but in larger specimens, the aboral spines are evenly spaced. *Leptasterias coei* has one row of superomarginals and one row of inferomarginals. It has two rows of oral intermediates at the base of the arm but one row reaches only to the middle of the arm. The proximal adambulacrals have one spine to a plate, but the middle to distal plates have one and two spines alternating. Small clusters of crossed and straight pedicellariae occur on the adambulacral spines. The adoral carina consists of three plates. There is a heavy apical spine on the oral plate with a smaller companion beside it.

This species was not included in Flowers' (1999) investigation as he did not obtain genetic information.

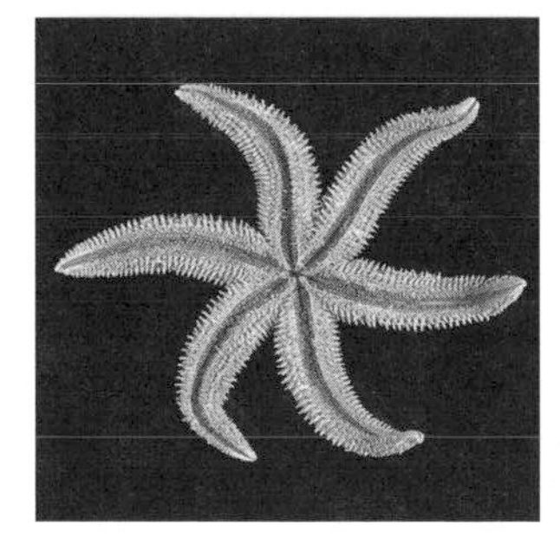

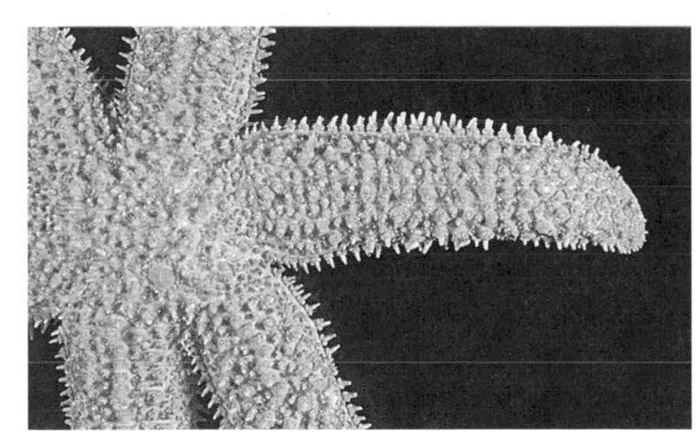

106. *Leptasterias coei*, R = 85 mm. (See colour photograph C-30.)

Similar Species

Leptasterias coei is superficially similar to *Orthasterias koehleri*, but the different colour and number of arms separate them. *L. coei* is also similar to *L. polaris katherinae*, which also has six arms, but comparing the arm-to-disc ratios and details of the mouth plates and adambulacral spines will distinguish them.

Distribution

Kodiak Island to Juneau, Alaska in depths of 18 to 187 metres on blue mud or fine sand. Uncommon.

Biology

There is no biological information on this poorly known species.

References

Fisher 1930, Verrill 1914. **Range:** Fisher 1930.

Leptasterias polaris katherinae

polaris:	**Latin *polus*, pole of the earth.**
katherinae:	**after Lady Katherine Douglas, who presented specimens to the British Museum in 1838.**

Description

Leptasterias polaris katherinae is a large six-armed sea star with an arm length of up to 14.8 cm and an arm-to-disc ratio of 4.5 to 5.3 It's well-spaced, subcapitate aboral spines do not form a clear carinal row. *L.p. katherinae* has a single row of superomarginal spines, one row of inferomarginals and two rows of oral intermediates. The adambulacral spines normally alternate one and two, with the furrow spine of the pair more slender than the distal one. The superomarginals have a complete wreath of pedicellaria; the inferomarginals and oral intermediates have a half wreath on the distal side. Each mouth plate bears three spines, a small spine guarding the entrance to the furrow, an intermediate flat spine pointing over the mouth, and a much longer tapered suboral spine.

Taxonomic Note: Originally described as *Asterias katherinae* Grey; revised to *Leptasterias polaris katherinae* by Fisher (1930).

Similar Species

Leptasterias polaris katherinae is similar to *L. coei*, which also has single superomarginal and inferomarginal spines, but the more robust aboral spines of the latter tend to be in regular rows and have a heavier wreath of pedicellaria around them. *L.p. katherinae* has an arm-to-disc ratio of 4.9 to 5.3 while *L. coei* has a ratio of 6.5.

107. *Leptasterias polaris katherinae*, R = 150 mm. (See colour photograph C-31.)

Apart from the general size and shape, the differences in details between the two are subtle and not easily determined.

Distribution

The species *Leptasterias polaris* ranges across the Arctic Ocean from the intertidal zone to 137 metres deep. There are four subspecies: *L.p. borealis* in the waters off eastern North America; *L.p. polaris* in the high Arctic of eastern North America; *L.p. acervata* from the Arctic to the Shumagin Islands and just south of the Alaska Peninsula to Kodiak Island; and *L.p. katherinae,* first described from the mouth of the Columbia River and in the Strait of Georgia, but it has not been documented from this region since. I have tentatively identified some specimens of *L.p. katherinae* from Glacier Bay and near Juneau at depths of 5 to 10 metres, but more records are needed to clarify its correct depth distribution.

Biology

Unknown, but the gonopore opens to the ventral side, like other *Leptasterias,* suggesting a brooding behaviour.

Reference

Fisher (1930).

Lethasterias nanimensis

nanimensis: **from the type locality: Departure Bay, Nanaimo, B.C.**

Description

Lethasterias nanimensis has five long flexible arms with numerous, black-tipped spines wreathed with crossed pedicellariae. The background colour on the upper surface is yellowish-brown. In alcohol, the black colour in the spines may be lost. The arms up to 30 cm long and the arm-to-disc ratio is 8. The aboral spines form a regular carinal series, but other spines are less regular. The tips of the spines tend to be fluted like a drill bit. The crossed pedicellariae have teeth of equal size. Occasional straight pedicellariae occur on the oral surface near the mouth plates. The marginals consist of a row of superomarginals with one spine (usually blackish) and a row of inferomarginals with two spines, the latter usually gouge shaped and bearing a cluster of pedicellariae on the distal side. Small clusters of up to ten papulae occur between the spines, especially near the marginals. The smaller adambulacrals (33 plates per 10 inferomarginals) each bear two spines, the proximal one smaller. The mouth plates are proximal to a short adoral carina of two fused adambulacrals and they each bear two marginal spines, one longer than the other, and one or two suborals.

The subspecies *L.n. chelifera* has more numerous straight pedicellariae on all parts of the body, with two or three longer curved interlocking fingers. The aboral spines are less regular and smaller.

Taxonomic Note: Originally described as *Asterias nanimensis* Verrill; revised to *Lethasterias nanimensis* by Fisher (1923).

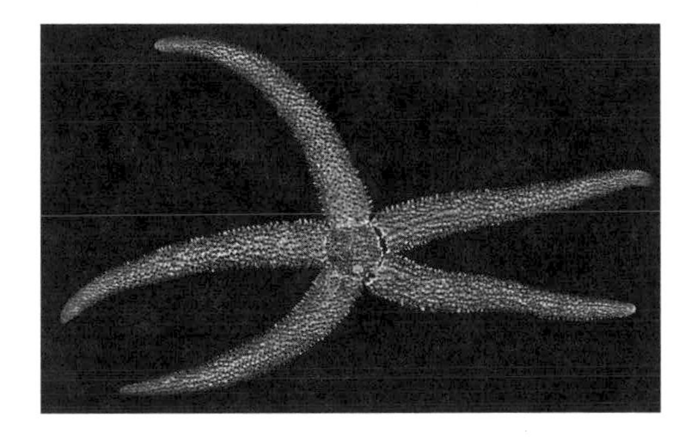

108. *Lethasterias nanimensis*, R = 110 mm. (See colour photograph C-32.)

Similar Species

Lethasterias nanimensis might be confused with *Orthasterias koehleri* or *Stylasterias forreri*, but the living colours are quite distinct. *O. koehleri* has a row of oral intermediates and two rows of inferomarginals. The crossed pedicellariae of *S. forreri* have distinctive "canine teeth".

Distribution

Southeast Alaska. Found in the shallow subtidal zone to 102 metres on mud, sand, gravel or boulders. I observed one specimen of *Lethasterias nanimensis* on a sand slope near Juneau. The species account for *L. nanimensis* was based on two specimens. The type specimen ostensibly from Departure Bay in Nanaimo, B.C., and another collected by the steamer *Albatross* in Juan de Fuca Strait. Fisher (1928) has doubts about the type locality data, and in 25 years of diving I have never seen this species in B.C. I identified museum specimens from around Juneau as *L. nanimensis*. Others have been reported farther south in Stephen's Passage, but its existence in northern B.C. has yet to be confirmed.

The subspecies *L.n. chelifera* ranges from the Bering Strait to the Sea of Japan and to Kodiak Island, Gulf of Alaska, from low tide to 224 metres deep. Fisher (1928) points out that the subspecies is the more important boreal form, ranging from the Bering Sea to the Gulf of Alaska, so we have the odd situation of the subspecies being widespread and the typical species having a limited distribution in Southeast Alaska.

Biology

The digestive glands are long and voluminous, reaching nearly to the tips of the arms. The gonopores occur at the arm angle near the superomarginal plates. The spacious ventral stomach has very strong retractors. When handled this species readily drops its arms. Bob Stone at the Auke Bay Marine Lab reported seeing hundreds of individuals on a shallow cobble bottom where the sea cucumber *Cucumaria frondosa japonica* was common. A month later none could be found at that spot.

References

Barr and Barr 1983, Lambert 1998. **Range:** D'Yakonov 1950, Fisher 1928.

Orthasterias koehleri — Rainbow Star, Long-armed Sea Star

koehleri: **after the French taxonomist R. Koehler.**

Description

Orthasterias koehleri is a strikingly colourful large long-armed sea star with reddish banding around the arms between white or cream patches and prominent white or purple spines. Some individuals are a plain straw colour or a shade of blue. *O. koehleri* has five arms up to 25 cm long and an arm-to-disc ratio of 6.5 to 10.0. The arms have a row of carinal spines, each 4 to 5 mm long, with two or three series of similar well-spaced rows of spines on each side of the carinals. The aboral spines have a wreath of crossed pedicellariae at the base. Lanceolate pedicellariae, which vary from straight and pointed to broad and toothed, occur between the spines. The papulae occur in groups of three to five (figure 110). A regular series of 80 to 85 superomarginals, with spines similar to aborals, curves upward in the interradial region to meet the series from the adjacent arm. Each spine has a wreath of crossed pedicellariae. An obvious intermarginal channel, with occasional lanceolate pedicellariae, separates this series from the two series of stout

109. *Orthasterias koehleri*, R = 240 mm. (See colour photograph C-33.)

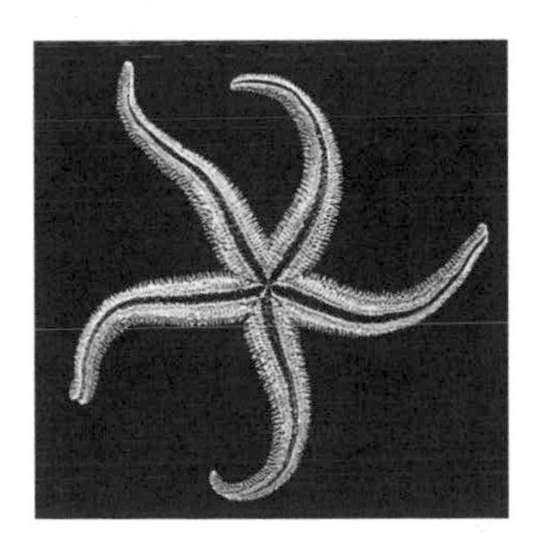

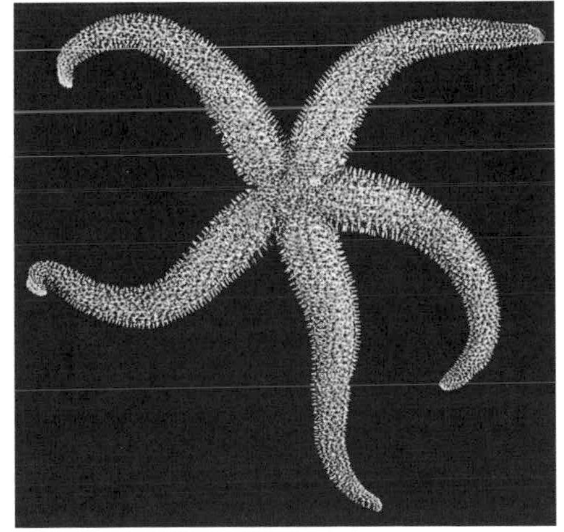

inferomarginal spines each with a tuft of crossed pedicellariae on distal side only (figure 111). Oral intermediates consist of a single row of stout spines between the inferomarginals and adambulacrals (figure 111). The adambulacrals have two slender diverging spines at right angles to the furrow, with a lanceolate pedicellaria at the base (figure 111). Three to five adambulacral plates fuse to form an adoral carina. The mouth plates are usually too sunken to be seen clearly (figure 112).

Taxonomic Note: Originally described as *Asterias koehleri* de Loriol; revised to *Orthasterias koehleri* by Verrill (1914).

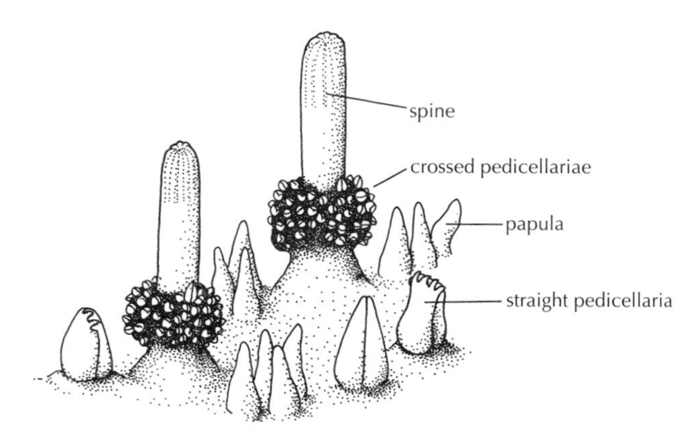

110. Aboral surface of *O. koehleri*.

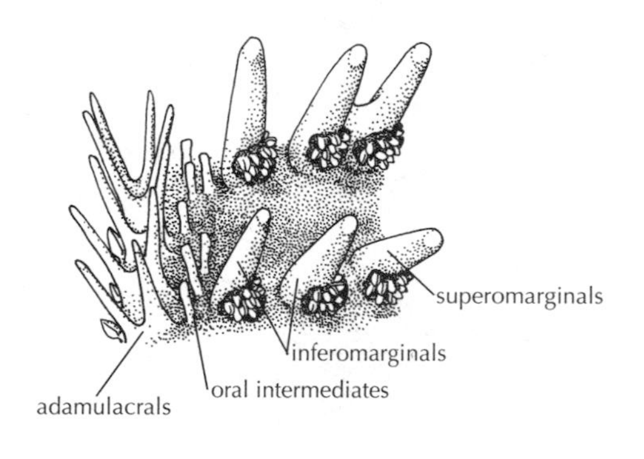

111. Oral arm spines.

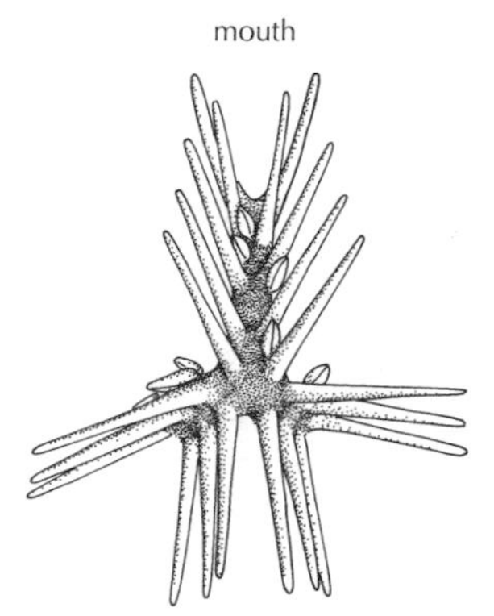

112. Adoral carina.

Similar Species

Orthasterias koehleri is similar to *Stylasterias forreri* in the aboral spines and wreaths of pedicellariae, but the living colours distinguish them. *S. forreri* is only black or brown, never red, orange or white. *O. koehleri* has a row of oral intermediates and *S. forreri* does not. *Lethasterias nanimensis* also has prominent aboral spines with a wreath of pedicellariae, but is usually straw-coloured with black-tipped spines.

Distribution

The eastern Aleutian Islands, Alaska, to Santa Rosa Island, California. Common on rock, pebbles, sand and mud in the intertidal zone to 230 metres deep.

Biology

In a study of 157 feeding individuals, 71 per cent ate the bivalve *Humilaria,* 10 per cent the rock jingle *Pododesmus,* 6 per cent the chiton *Tonicella,* 3 per cent the lamp shell *Terebratalia,* and 2 per cent the black chiton *Katharina.* Other feeding observations include sea squirts and bivalves (*Entodesma, Hinnites* and *Saxidomus*).

The species breeds from June to August. It elevates itself on arm tips to spawn from paired gonopores. Females produce translucent tangerine-coloured eggs (diameter 150 micrometres). After fertilization, these develop into planktotrophic bipinnaria larvae in 5 days at 10°C. Ripe female *O. koehleri* produce an unidentified chemical that attracts sperm. This may be a general phenomenon with other sea stars, but it has only been demonstrated with this species. The scale worm *Arctonöe fragilis* is commensal on *O. koehleri.*

Richard Carlson at the Auke Bay Lab in Juneau reported that a tagged individual that had lost an arm had regenerated only three quarters of it after six years, demonstrating that growth is quite slow. He also found that individuals live at least nine years and probably longer.

References

Carlson in O'Clair and O'Clair 1998, Christen 1985, Hopkins and Crozier 1966, Mauzey et al. 1968, Miller 1989, Stickle 1985c, Stricker et al. 1994a, Webster 1975. **Range:** Barr and Barr 1983, Fisher 1928.

Pisaster brevispinus **Giant Pink Star**

***brevispinus*: Latin *brevis*, short, and *spina*, thorn.**

Description

Pisaster brevispinus is a large stiff-armed, pinkish-grey sea star usually found on soft substrates. It has five arms up to 32 cm long and an arm-to-disc ratio of 2.8 to 5.0. The aboral surface is a meshwork of plates with smallish spines standing alone or in clusters, with crossed and furcate pedicellariae in wreaths around the spines or standing on their own (figure 114). Abundant grey or light purple papulae occur between the spines, interspersed with large lanceolate pedicellariae (figure 114). The superomarginal spines are larger than the aboral spines and are separated from the inferomarginals by an obvious channel. Two or three stout inferomarginal spines have clusters of crossed and furcate pedicellariae on their distal sides. Two series of oral intermediates with stout spines also have clusters of pedicellariae on their distal sides. The adambulacrals bear a single slender spine with a tuft of pedicellariae on the plate near the base of the spine. Twelve to fifteen adambulacral plates form a long adoral carina, with mouth plates at the proximal end usually sunken and obscured by tube feet.

Taxonomic Note: Originally described as *Asterias brevispina* Stimpson; revised to *Pisaster brevispinus* by Verrill (1909).

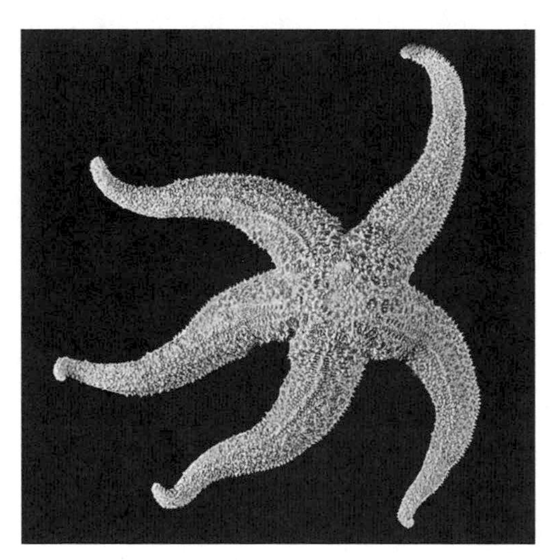

113. *Pisaster brevispinus*, R = 230 mm. (See colour photograph C-36.)

Similar Species

No other five-armed sea star in this region can match *Pisaster brevispinus* for size and bulk. Its smaller relative *P. ochraceus* is never pinkish-grey and is usually found in the lower intertidal zone of rocky shores, not on soft substrates. A couple of historical records report *P. giganteus* from southern Vancouver Island, but I have not been able to confirm its presence.

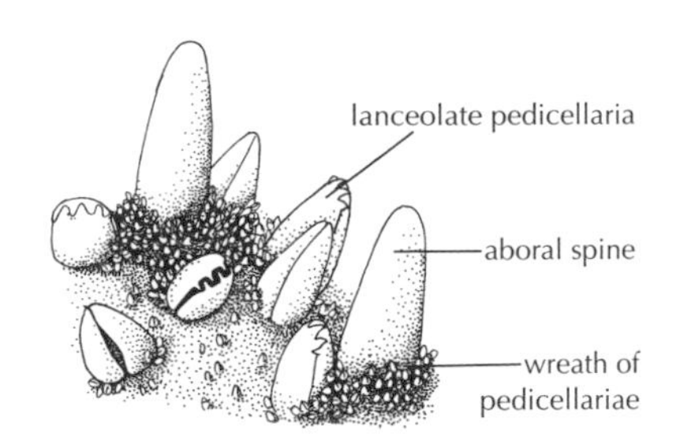

114. Aboral surface of *P. brevispinus.*

Distribution

Sitka, Alaska, to Santa Barbara, California, from the intertidal zone to 128 metres deep. Common in shallow waters on mud, sand or broken shell. Smaller individuals also occur on rocky substrates.

Biology

Pisaster brevispinus captures bivalves by elongating the tube feet near the mouth to penetrate the substrate and grasp prey, such as the Butter Clam (*Saxidomus giganteus*), Little Neck Clam (*Protothaca staminea*), Geoduck (*Panope generosa*) and Ribbed Clam (*Humilaria kennerleyi*). *P. brevispinus* attacks rock boring clams by inserting its tube feet, drawing the clam towards the mouth and applying its stomach. In winter, 70 per cent of observed *P. brevispinus* were feeding, with 70 per cent of these on barnacles and the rest on Geoducks and Jackknife Clams (*Solen sicarius*). In California, *P. brevispinus* feeds on the Horse Clam (*Tresus capax*) and the Eccentric Sand Dollar (*Dendraster excentricus*) which tries to escape by burying itself in the sand. In Puget Sound, however, *P. brevispinus* shows little preference for sand dollars. On rocky substrates it eats the Giant Barnacle (*Balanus nubilus*). The Olive Snail (*Olivella biplicata*) exhibits a strong escape response when approached by *P. brevispinus*.

Fisher describes *P. brevispinus* as one of the largest known sea stars. It breeds from April to August. Divers have seen it spawning in August with its central disc humped up and gametes streaming from paired gonopores. In December, nutrient stores in the gonad increase and the nutrients in the pyloric caeca diminish.

References

Carey 1972, Cool et al. 1988, Edwards 1969, Farmanfarmaian et al. 1958, Haderlie 1980, Hopkins and Crozier 1966, Kovesdi and Smith 1985, Mauzey et al. 1968, Pearse et al. 1988, Sloan and Robinson 1983, L.S. Smith 1961, M.J. Smith 1982, M.J. Smith et al. 1982, Van Veldhuizen and Oakes 1981, Van Veldhuizen and Phillips 1978, Webster 1975, Wobber 1975, Zollo et al. 1989. **Range:** Fisher 1930, Lambert 1999.

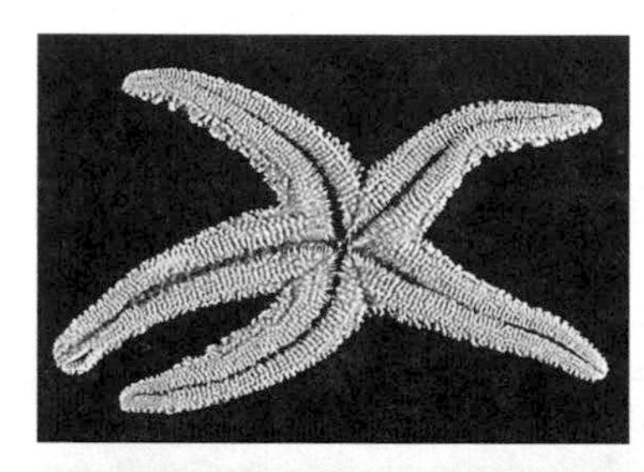

115. *Pisaster ochraceus*, R = 122 mm. (See colour photographs C-34, C-35 and on the back cover.)

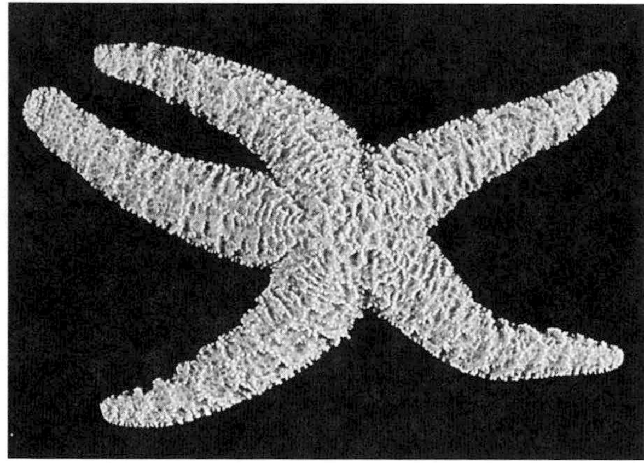

Pisaster ochraceus — Purple Star, Ochre Star

ochraceus: **Greek *ochros*, pale yellow.**

Description

Pisaster ochraceus is our most common intertidal sea star, ranging in colour from purple to orange or yellow, with white spines and white to pale yellow tube feet. It has five stiff arms, a highly arched disc and a sunken mouth, and attaches strongly to the rocks at low tide. The arms are up to 25 cm long (an RBCM specimen) and the arm-to-disc ratio ranges from 2.7 to 4.1. The aboral surface has an extremely variable pattern of spines ranging from small numerous spines in a netlike pattern, to stout knobby spines in well-separated convex groups. The papulae, between aboral spines, give the sea star its basic colour. Among the papulae are furcate, crossed and small lanceolate pedicellariae that can only be seen when magnified (figures 4 and 116), and occasionally several larger straight toothed pedicellariae. The furcate pedicellariae are characteristic of this species. On lower side of the arm, superomarginals form a regular row with an intermarginal channel just below. Between the intermarginal channel and the adambulacrals are five to eight rows of similar spines. The first two rows are inferomarginals, the next three to six are oral intermediates. Each spine has a tuft of pedicellariae on the distal side. The adambulacrals have one slender spine per plate and small lanceolate pedicellariae attached to the plate but not the spines. Ten adambulacrals fuse together as an adoral carina. The mouth plates are sunken and difficult to see, but each plate has two stout marginal spines.

Taxonomic Note: Originally described as *Asterias ochracea* Brandt; revised to *Pisaster ochraceus* by Fisher (1908).

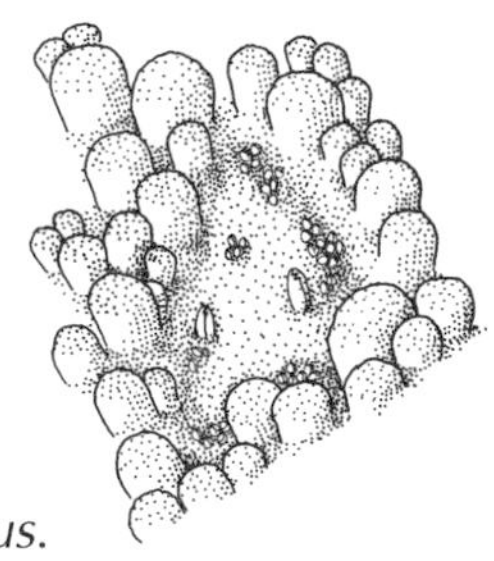

116. Aboral surface and furcate pedicellaria (above) of *P. ochraceus*.

Similar Species

Pisaster ochraceus might be confused with *Evasterias troschelii* (the Mottled Star), but the latter has a mottled colour, a longer arm-to-disc ratio, a shallow mouth and, alternately, one and two spines on the adambulacrals rather than one per plate.

Distribution

Prince William Sound, Alaska, to Cedros Island, Baja California, from the intertidal zone to 97 metres deep. Abundant on rocky shores in middle to lower intertidal zone exposed to waves or currents. On more sheltered parts of the coast, such as mainland inlets, this species tends to be replaced by *Evasterias troschelii*.

Biology

Pisaster ochraceus feeds primarily in summer, preferring mussels, barnacles, limpets and snails. About 30 prey items have been documented but diet depends on the availability of prey. In the presence of *P. ochraceus* large Turban Snails (*Tegula funebralis*) and the limpets *Acmaea limatula* and *Acmaea scutum* escape by moving up vertical surfaces to higher places on the shore. *Margarites* species, *Calliostoma ligatum* and *Amphissa columbiana* show a strong escape reaction. *P. ochraceus* eats mussels by inserting its stomach between the shells, secreting digestive enzymes and, at the same time, pulling the shells apart with its tube feet. In Puget Sound in winter, *P. ochraceus* migrates lower on the shore and stops feeding. Near Monterey, California, an average-sized sea star eats about 80 mussels per year and does not show a marked summer feeding or downward migration in winter. The clam *Chama arcana* is more abundant in patches of Strawberry Anemones (*Corynactis californica*) because *P. ochraceus* is

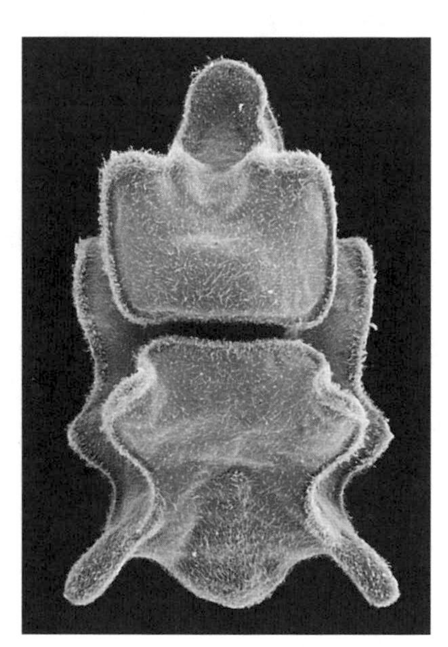

117. *P. ochraceus* larva: bipinnaria. (T.H.J. Gilmour and T.C. Lacalli photograph.)

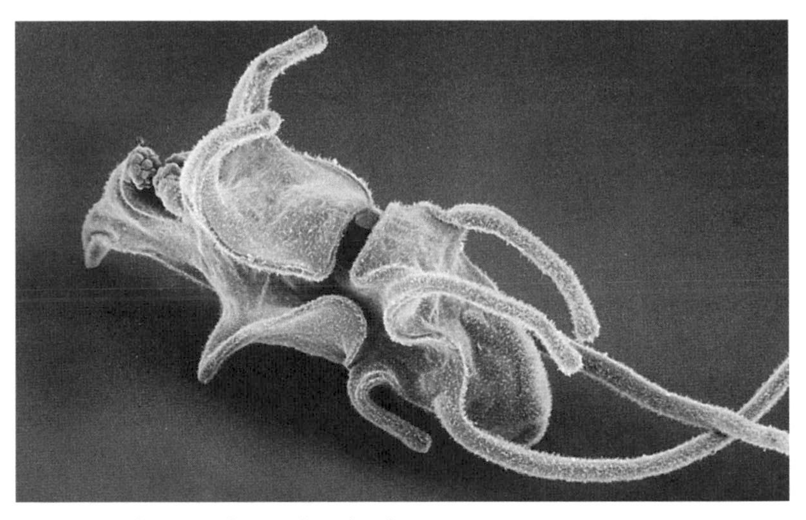

118. *P. ochraceus* larva: brachiolaria.
(T.H.J. Gilmour and T.C. Lacalli photograph.)

deterred by the stinging cells of the anemone. If a number of *P. ochraceus* are placed on the shore in a group they tend to move away from each other as quickly as they can. This behaviour ensures that they are more evenly dispersed in the habitat and reduces competition for food. Their role as a keystone predator is discussed in papers by Paine (1984 and 1985).

The gonad starts to grow in January, in preparation for spawning from May to July, when it produces pale orange to salmon-pink eggs (diameter 150 to 175 micrometres). A 400-gram sea star can release 40 million eggs. The length of day and night (photoperiod) has been shown to control the onset of gamete production and gonad growth. A bipinnaria larva (figure 117) forms after about six days and can survive up to two months in the plankton. Towards the end of this planktonic phase, it develops into a brachiolaria larva (figure 118). The brachiolaria has sticky arms and a sucker for attaching to the substrate, where it will metamorphose into a juvenile sea star. Larvae of this species use a chemical defence that causes filter feeders like the mussel *Mytilus edulis* and the sea squirt *Styela gibbsii* to reject a significant proportion of them. The larvae can tolerate a salinity as low as 20 parts per thousand, but many are abnormal and smaller. Young stars reach maturity at 5 years and live for 20 years or more. The larval nervous system is described in Burke 1983 and Lacalli et al. 1990. The scale worm *Arctonöe fragilis* is commensal on *P. ochraceus*. A parasitic ciliate *Orchitophrya stellarum*, endemic to the North Atlantic, was discovered in 1988 in the Purple Sea Star in British Columbia. It infests the male gonad causing castration and loss of sperm production. This same parasite was reported in *Asterias amurensis* in Japanese waters in 1990. The only predators known to prey on adult *P. ochraceus* are sea otters and sea gulls.

P. ochraceus is common and easy to collect and thus has been used for numerous general studies in physiology, embryology and pharmacology. Many of those papers are listed below.

References

Banfield et al. 1988, Boom and Smith 1989, Burke 1983, Burke and Watkins 1991, Byrne et al. 1998, Campbell and Crawford 1991, Cloud 1983 and 1984, Cool et al. 1988, Cowden et al. 1984, Crawford 1988, 1989 and 1990, Crawford and Abed 1983, Crawford and Chia 1978, 1980 and 1982, Crawford and Crawford 1992, Daya-Makin et al. 1991, Desantis and Cloud 1984 and 1988, Doering and Phillips 1983, Farmanfarmaian et al. 1958, Farrand and Williams 1988, Fawcett 1984, Feder 1955, 1959, 1963 and 1970, Ferguson 1992 and 1994, Fraser et al. 1981, Garnas and Crosby 1979, Halberg et al. 1987, Hoffman 1980, Holzman et al. 1985, Hopkins and Crozier 1966, Howell et al. 1987, Jacobs et al. 1989, Johnson and Epel 1982, Jonas-Davies and Liston 1985, Kent 1981, Kovesdi and Smith 1982 and 1985, Kunz and Conner 1986, Lacalli 1993, Lacalli et al. 1990, Lacalli and West 1993, LeClair 1993, Leighton et al. 1991, Mauzey et al. 1968, Meijer et al. 1987, Menge 1975, Menge et al. 1994, Menge and Menge 1974, Mortensen 1921, Moss et al. 1994, Obrietan et al. 1991, Otto and Schroeder 1984, Paine 1984, Paine et al. 1985, Palumbi and Freed 1988, Patton et al. 1991, Pearse and Eernisse 1982, Pelech 1995, Pelech et al. 1987 and 1988, Phillips 1975, Quayle 1954, Reimer and Crawford 1990, Reimer et al. 1992, Rice 1985, Roller 1988, Roller and Stickle 1985, Sanghera et al. 1990 and 1991, Schroeder and Stricker 1983, Sewell and Watson 1993, Smith 1982, Smith et al. 1989 and 1990, Smith and Boal 1978, Smith et al. 1982, Stickle et al. 1992, Strathmann 1971, Stricker et al. 1994a and 1994b, Stricker and Schatten 1991, Taylor and Williams 1995, Van Veldhuizen and Oakes 1981, Zollo et al. 1989.
Range: Barr and Barr 1983, Maluf 1988.

Pycnopodia helianthoides **Sunflower Star**

***helianthoides*: Greek *helianthes*, sunflower, and *eides*, like.**

Description

Pycnopodia helianthoides is the largest sea star. It has more arms (15 to 24) than any other species and is softer than most. It is also probably the heaviest known sea star, weighing in at about 5 kg. The colour varies from reddish-orange to yellow, violet brown, purplish or slatey purple – the colour can depend on how much of the underlying skin is exposed when the papulae or pedicellariae expand. The oral surface is usually lighter, with yellow to orange tube feet. The arms are up to 40 cm long and the arm-to-disc ratio is 2.5 to 3.5. The aboral surface is soft and flexible, because the calcareous plates are not connected to one another. The plates near the centre of the disc and the base of arms sometimes bear stubby spines surrounded by a wreath of crossed pedicellariae and occasional lanceolate pedicellariae. The number of spines on the arms decline distally and the clumps of pedicellariae are closer together, with up to 75 papulae scattered among them. A slight furrow devoid of papulae occurs from where the arms join to the aboral surface. The superomarginals make up the first row of prominent spines on the side of each arm. Below these are the inferomarginal plates, bearing two rows of spines that overshadow the insignificant adambulacrals with one small spine. Three adambulacral plates are about equal to one inferomarginal. At the base of each adambulacral spine is a cluster of small straight pedicellariae that are usually hidden by the large numerous tube feet. Ten to fifteen adambulacrals form an adoral carina. The mouth plates each have two apical spines and one suboral spine, and clusters of straight pedicellariae.

Taxonomic Note: Originally described as *Asterias helianthoides* Brandt; revised to *Pycnopodia helianthoides* by Stimpson (1862).

Similar Species

A juvenile *Pycnopodia helianthoides* might be mistaken for a *Crossaster papposus*, but the details of their aboral surfaces readily distinguish the two. Deeper than 200 metres, it can be confused with *Rathbunaster californicus*.

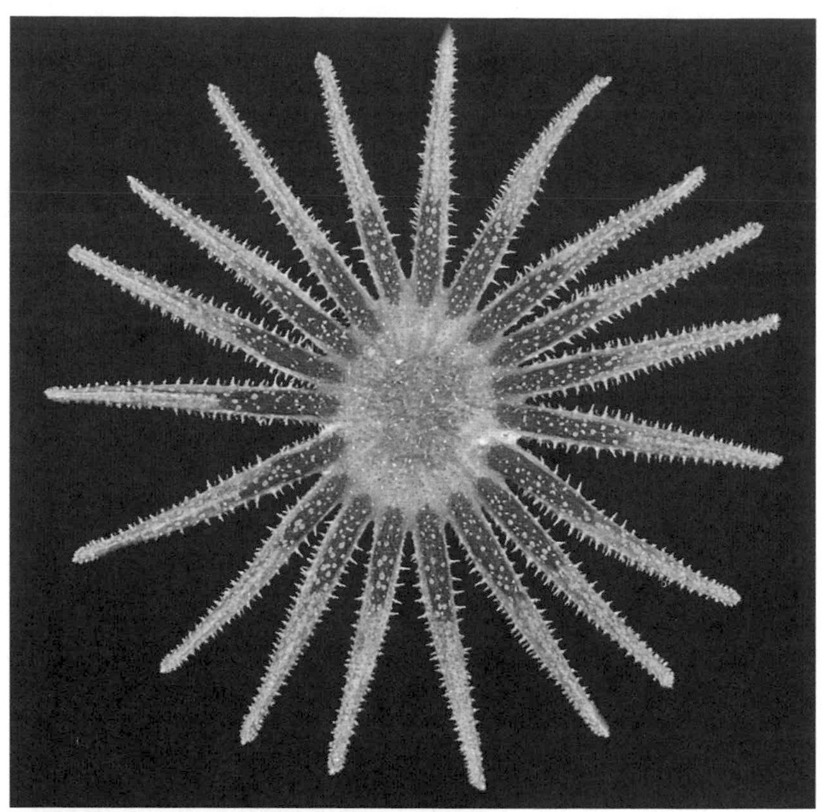

119. *Pycnopodia helianthoides*, R = 147 mm. (See colour photograph C-37.)

Distribution

Unalaska, Aleutian Islands, to San Diego, California. Common to abundant on a variety of substrates, such as mud, sand, gravel, boulders and rock, from the intertidal zone to 435 metres deep. But only one specimen was from 435 metres; all the others in Fisher 1928 and those that I have examined in museum collections were from depths of no more than 120 metres, so the 435-metre depth is abnormal and possibly an error.

Biology

The diet of *Pycnopodia helianthoides* varies with geographic location and the availability of prey, such as sea urchins, hermit crabs, sea cucumbers, clams and sand dollars. In the San Juan Islands, its main diet is the clam *Saxidomus* but on the exposed west coast, one study showed sea urchins as the main prey. *P. helianthoides* captures clams by excavating the gravel piece by piece with its arms and tube feet until it exposes the clam, then engulfing it with the everted stomach. The Sunflower Star swallows sea urchins whole. In food preference tests, it was equally attracted to Purple and Red sea urchins (*Strongylocentrotus purpuratus* and *S. fransiscanus*), but chose Purple Sea Urchins 90 per cent of the time. A study in Barkley Sound found 72 per cent of Sunflower Stars in the act of feeding. Of these, most took snails and clams, followed by crustaceans, and only 4.6 per cent ate Purple Sea Urchins. Larger Sunflower Stars have been found in areas of intermediate exposure on soft substrates, with juveniles more common in protected waters. In the winter in Gabriola Passage, 16.6 per cent were feeding and 80 per cent of those were eating barnacles. In Prince William Sound, Alaska, *P. helianthoides* feeds on molluscs, but in Glacier Bay, it commonly eats Green Sea Urchins (*S. droebachiensis*) and Purple Sea Urchins. Both urchins show strong escape responses. Much of *P. helianthoides'* nutrient reserves are stored in the body wall as well as in the pyloric caeca.

P. helianthoides can outrun most benthic species, travelling up to 160 cm per minute. To defend against this predator, some species have developed specific escape responses. The abalone *Haliotis* accelerates and simultaneously whips its shell back and forth to break the grasp of the sea star's tube feet. The snails *Tegula brunnea* and *Calliostoma ligatum* show typical flight responses as well as shell twisting. The swimming scallops *Chlamys* spp., the California

Sea Cucumber (*Parastichopus californicus*), the nudibranch *Dendronotus iris* and the anemone *Stomphia didemon* swim away when touched by this sea star.

P. helianthoides breeds from March to July by broadcast fertilization. Eggs (120 micrometres in diameter) develop into swimming larvae that may remain in the plankton for up to ten weeks feeding on single-celled plants. After the larva has settled to the bottom and metamorphosed into a young sea star, it feeds initially on the thin layer of single-celled plants coating the bottom. Near the Auke Bay Lab, Richard Carlson tagged two specimens and later saw them 3 km away.

The Sunflower Star readily drops off (autotomizes) its arms when disturbed or irritated as a result of the rapid change in the tensile strength of the catch connective tissue, described in the introduction. Autotomy is triggered by a chemical that is released by injured tissues. Presumably it allows the Sunflower Star to escape from a predator holding onto its arm. Juices from the injured sea star cause an alarm response (increased movement in the direction of the current) in nearby Sunflower Stars.

References

Ahearn 1990, Ahearn and Behnke 1991, Alexander and Dresden 1980, Asotra et al. 1988, Bavendam 1985, Birkeland et al. 1982, Bruno et al. 1989, Caine and Burke 1985, Cavey and Wood 1988, Duggins 1983, Feder 1963, Fisher 1928, Francour 1997, Greer 1962, Hoffman 1980, Kjerskog-Agersborg 1918, Kunz and Conner 1986, Lambert 1981b and 1988, Lawrence 1991, McClintock 1989, Mladenov et al. 1989b, Moitoza and Phillips 1979, Mortensen 1921, Nance and Braithwaite 1979, Paul and Feder 1975, Pearse and Hines 1987, Phillips 1978, Punnett et al. 1992, Rumrill 1989, Schiel and Welden 1987, Sewell and Watson 1993, Shivji et al. 1983, Sloan and Robinson 1983, Smith 1982, Strathmann 1971, Van Veldhuizen and Oakes 1981, Watanabe 1983, Webster 1975, Wobber 1975.
Range: Fisher 1928.

Stylasterias forreri **Long-rayed Star**

forreri: **after Alphonse Forrer, who collected this species off Santa Cruz, California.**

Description

Stylasterias forreri is a large slender-armed, black sea star with large aboral spines surrounded by prominent wreaths of pedicellariae. The common large form has white spines, grey wreaths of pedicellariae, and black skin and papulae with yellow tube feet on the lighter oral side. Other colour morphs, usually found on rock, are grayish-brown and, rarely, a straw colour. The five arms reach up to 33.5 cm in length and the arm-to-disc ratio is 11 to 16. On each arm long (4 to 5 mm) slender white spines occur in three to five longitudinal series. Each spine has a heavy wreath of crossed pedicellariae (figures 120 and 123). Between the wreathed spines are groups of 8 to 10 papulae with an occasional large straight pedicellaria. On the side of the arm, the superomarginals bear a closely spaced, regular row of pointed spines surrounded by a smaller wreath of pedicellariae than the aboral spines. The inferomarginals have two scoop-shaped spines with a tuft of pedicellariae on the aboral side of each spine. The adambulacrals bear one thin furrow spine and another slightly larger flat spine on the oral surface (figure 121). Small mouth plates, proximal to a short adoral carina of two plates, have two or three marginal spines and one suboral.

Taxonomic Note: Originally described as *Asterias forreri* de Loriol; revised to *Stylasterias forreri* by Fisher (1928).

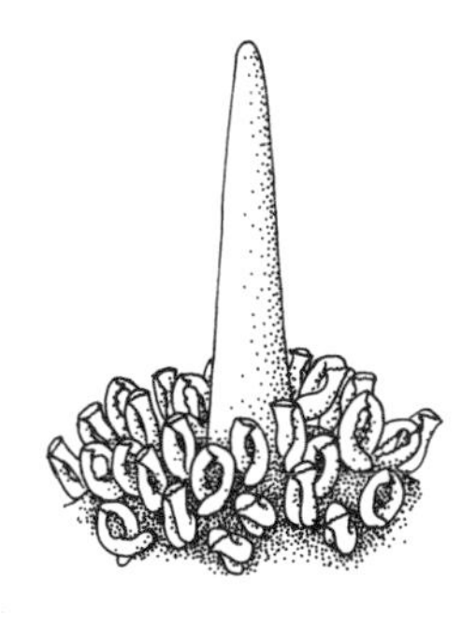

120. Crossed pedicellariae surrounding an aboral spine.

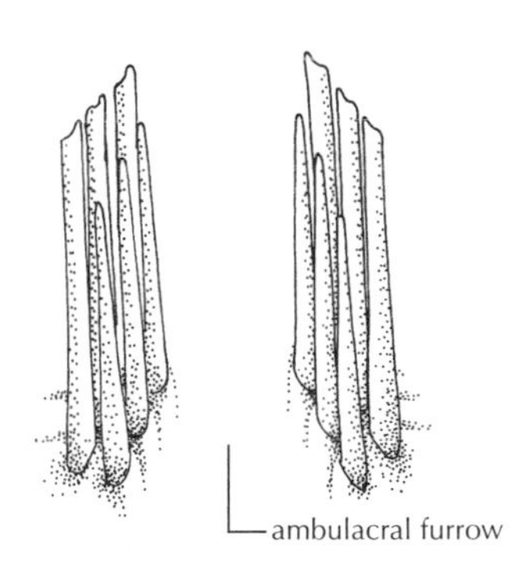

121. Adambulacral spines of *S. forreri*.

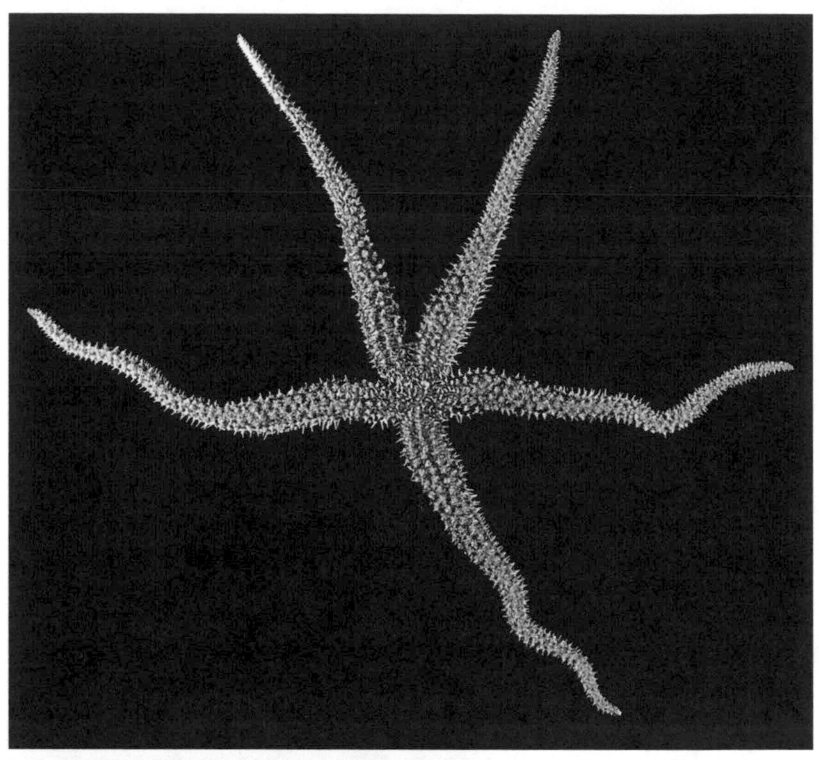

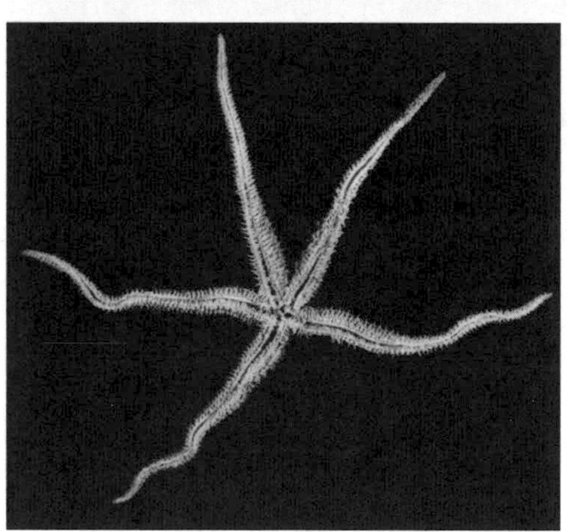

122. *Stylasterias forreri*, R = 335 mm. (See colour photographs C-38 and C-39.)

Similar Species

Stylasterias forreri might be confused with *Lethasterias nanimensis* or *Orthasterias koehleri*, but the living colours are quite distinctive. Otherwise, the details of the aboral pedicellariae should be examined.

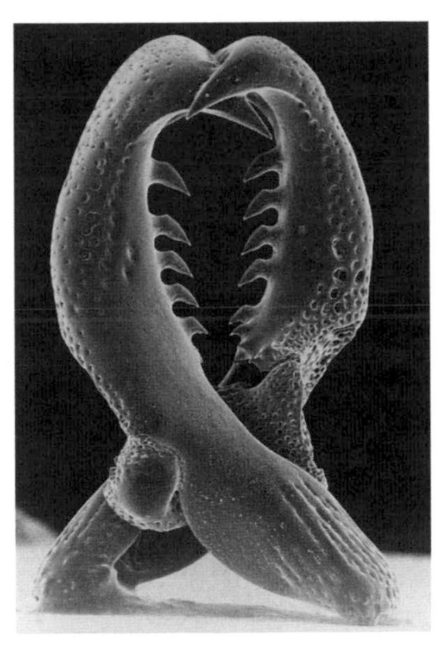

123. Scanning electron micrograph of a crossed pedicellaria (1.6 mm tall) of *S. forreri.* (Fu-Shiang Chia photograph.)

Distribution

Kodiak Island to San Diego, California; Uncommon, found on rock and shell-gravel at depths of 6 to 532 metres.

Biology

Stylasterias forreri feeds primarily on snails (*Thais lamellosa*, *Calliostoma ligatum* and *Margarites* spp.) and chitons (*Tonicella* spp.), using normal prey-capture methods. In addition, the wreaths of toothed pedicellariae can snare small sculpins and scallops that accidentally settle on the aboral surface, which are then transferred to the mouth. It can move at 9 to 32 cm per minute. Six *S. forreri* were observed spawning in August while elevated on the tips of their arms. Gametes were shed from gonopores between each arm. Spawning seemed to be stimulated by the mass spawning of other species. The scale worm *Arctonöe fragilis* is commensal on *S. forreri.*

References

Carey 1972, Cavey and Wood 1981 and 1988, Chia and Amerongen 1975, Mauzey et al. 1968, McDaniel 1971, Pearse et al. 1988, Robilliard 1971, Wood and Cavey 1981. **Range:** Fisher 1928, Lambert 1999.

Family Pedicellasteridae

No species in this region.

Deeper than than 200 metres:
Ampheraster marianus
Pedicellaster magister
Tarsaster alaskanus

Family Labidiasteridae

No species in this region.

Deeper than than 200 metres:
Rathbunaster californicus

ORDER BRISINGIDA

Family Brisingidae

No species in this region.

Deeper than than 200 metres:
Astrocles actinodetus
Astrolirus panamensis
Craterobrisinga synaptoma

Family Freyellidae

No species in this region.

Deeper than than 200 metres:
Freyella microplax
Freyellaster fecundus

GLOSSARY

Abactinal (See aboral.)
Aboral The upper side of a sea star; the side opposite the mouth; also abactinal.
Aboral tabulate plates Plates shaped like a mushroom or a stout peg, with granules on the upper surface (figure 5).
Actinal See oral.
Actinolateral membrane A membrane supported by long spines that originate on the distal ends of the adambulacrals and extend out towards the edge of the arm. See *Pteraster* (figure 83).
Adambulacrals Calcareous plates that form the sides of the ambulacral furrow (figures 27 and 28).
Adoral carina A structure consisting of one to several adambulacrals from adjoining arms fusing together at the distal side of the mouth plates; typical of Asteriidae (figure 112).
Adradial nerve cord A smaller nerve that runs parallel to the radial nerve just beneath the ampullae of the tube feet.
Allele One of a number of different forms of a gene. Every offspring inherits two alleles for each gene, one allele from each parent.
Ambulacral furrow Furrow on the oral (underside) of each arm formed by pairs of ambulacral plates and from which many tube feet extend (figure 29).
Amino acids The basic structural units of protein.
Ampulla A saclike structure at the inner end of a tube foot that contracts to extend the foot by hydraulic pressure (figures 6 and 8).
Anus Posterior opening of the intestinal tract. In sea stars, the anus is usually in the centre of the aboral side of the disc.
Apical spines Spines located at the tip, or apex, of a structure, as in apical mouth spines.
Bilaterally symmetrical Having two equal halves.

Binomen A two-part scientific name for a species, consisting of the capitalized generic name followed by the lower-case specific name; e.g., *Henricia leviuscula.*

Bipinnaria larva The characteristic larval form of sea stars that bear ciliated bands and long, armlike projections (figures 9 and 117).

Brachiolaria larva A sea star larva with a sucker at the anterior end and three arms with adhesive tips (figures 9 and 118).

Brooder A female sea star that protects the fertilized eggs in a space around the mouth or in a special cavity beneath her body.

Bryozoa Microscopic invertebrates that form mosslike colonies on rocks and other substrates. They are packed together in boxlike cubicles of chitin (calcium carbonate) and feed with a spiral structure called a lophophore. Commonly called moss animals.

Calcareous plates Calcium carbonate elements that make up the internal skeleton of a sea star. The shape of the plates and their arrangement is characteristic for each species. Also called ossicles.

Capitate Refers to a spine with a caplike rounded tip that is broader than the base.

Cardiac stomach The lower of two stomachs in a sea star. It can be protruded through the mouth for external digestion (figure 7).

Carinals A row of plates running down the middle of the aboral surface of the arm, usually bearing obvious spines.

Cilia Movable, hairlike extensions of surface cells that move material across the surface or propel the animal through a fluid medium.

Circumboreal A region around the globe at the level of the north temperate seas.

Circumpolar A region around the globe at the level of the Arctic or Antarctic.

Coelom An internal body cavity that originates from a splitting or outpocketing (evagination) of the third embryonic layer (mesoderm). In sea stars, the coelem develops by outpocketing, and is filled with coelomic fluid that serves the function of a circulatory system.

Coelomocytes Cells freely suspended in coelomic fluid primarily concerned with defence against infection. Some coelomocytes can wander through tissues.

Commensal An organism that benefits from a host without harming or helping it.

Cribriform organs Specialized grooves, between marginal plates, lined with cilia that create a current to move food particles from the aboral surface to the mouth. (See *Ctenodiscus crispatus,* page 46.)

Dermal branchiae (See papulae.)

Detritus Debris or waste material. In this book it refers to organic material in bottom sediments resulting from the decomposition of animal and plant material.

Direct development Growth from egg to adult without undergoing an obvious radical metamorphosis in the larval stage.

Disc The central part of a sea star from which the arms extend (figure 3).
Distal In a direction or position away from the centre or central axis.
Endemic Restricted to a particular geographic region.
Foraminiferans Single-celled planktonic marine animals enclosed in a coiled or chambered shell that is usually composed of calcium carbonate.
Furrow spines Spines located on the inner edge of the adambulacral plates, overhanging or projecting into the ambulacral furrow (figure 29).
Gastropods Members of the phylum Gastropoda, which includes slugs and snails.
Gastrulation A stage in the development of the embryo in which a hollow ball of cells (the blastula) is indented with a pocket that later develops into the digestive tract.
Gonads Organs that produce eggs or sperm.
Gonopore The external opening through which eggs or sperm are released.
Granuliform spinelets Short stubby spinelets resembling granules.
Haemal system A set of vessels associated with various internal organs of a sea star.
Haplotype The linear, ordered set of alleles on a chromosome.
Holotype A single specimen deposited in a museum and used as the basis for the original scientific description.
Inferomarginals Plates forming the lower side of the arm between the oral and aboral plates (figure 27); also called infra, inferior or lower marginals.
Intermarginal channel A groove, devoid of plates or spines, situated between the superomarginals and inferomarginals.
Intermarginals One or more rows of plates lying between the superomarginals and inferomarginals (figure 27).
Interradial area The area between the radii of two arms.
Intertidal zone The portion of a shore that lies between the highest and lowest tides of the year.
Intestinal caeca Blind sacs that branch off the intestine between the pyloric stomach and the anus (figure 7); also called diverticula, rectal caecae or rectal sac. They help in compacting waste matter.
Lecithotrophic larva A nonfeeding larva that survives on an internal yolk supply.
Longitudinal series A row of spines that runs parallel to the ambulacral furrow, or some longitudinal axis (figure 29).
Madreporite A modified inter-radial plate on the aboral side of the disc pierced by numerous small pores that connect with the water-vascular system (figure 6).
Marginals Series of skeletal plates along the side of the arm between aboral and oral series; the term includes superomarginals, intermarginals and inferomarginals (figure 27).

Marginal spines Spines on the edge of a mouth plate (figure 28); may also refer to spines on marginal plates.

Metamorphosis In sea stars, the process by which a bilaterally symmetrical larva transforms into a radially symmetrical sea star.

Mouth plate The modified adambulacral plate closest to the mouth; two adjacent mouth plates form a triangle (figure 28).

Multidenticulate Possessing several points, as in a multidenticulate spine of *Henricia leviuscula*.

Nerve ring The nerve that encircles the mouth.

Nidamental chamber A space between the true aboral surface and a thin skin or supradorsal membrane supported by the tips of the pseudopaxillar spines of *Pteraster* or *Diplopteraster* (figure 80). In some species it acts as a brood chamber; nidamental is from the Latin *nidus*, meaning nest.

Oral The mouth side or underside of a sea star; also actinal.

Oral interradial area A triangular area on the oral side bordered by two ambulacral furrows and the edge of the disc (figure 28); also actinal interradial area.

Oral intermediates The series of plates that lie between the inferomarginals and the adambulacrals; also known as actinal intermediates. These plates may run the whole length of arm or may be restricted to the interradial area (figures 27 and 28).

Osculum A central pore in the supradorsal membrane of the Family Pterasteridae through which water is expelled from the nidamental chamber.

Papulae The gills of a sea star, thin-walled sacs that extend from the coelom to the outside, between the calcareous plates. Most papulae have a simple tubular or conical shape, but some are branched, as in *Luidia* and *Pycnopodia* (figure 8).

Papularia Well-defined round patches of papulae on the aboral surface of a sea star; typical of the Family Benthopectinidae.

Paxillae The columnar or mushroomlike aboral plates of some sea stars. The expanded tops are covered with moveable spinelets (figure 5). Singular: paxilla.

Paxilliform Shaped like a paxilla.

Pedicellariae Small jawlike structures that occur in numbers on the outer surface of many sea stars. Singular: pedicellaria. Seven kinds of pedicellariae are mentioned in this book:

- **Bivalve –** Two-jawed, shaped like a clam shell, as in the genus *Hippasteria* (figure 57).
- **Crossed –** The two parts of the jaw cross each other like the blades of scissors (figures 120 and 123).
- **Furcate –** A type of straight pedicellaria characteristic of *Pisaster* species (figures 4 and 116).

Lanceolate – Straight with two long narrow jaws resembling a spear head or lance (figure 114).

Pectinate – Made up of two sets of opposing teeth that interlock when closed (figures 4 and 39).

Spatulate – Straight with jaws that are flat at their tips so that they resemble a spatula.

Straight – Both parts of the jaw attach to the base and do not cross each other, so that it looks similar to forceps or tongs.

Pelagic Living or occurring in open water, and not associated with the sea bottom.

Phylogenetic Pertaining to the evolutionary origins of genetically related groups; from the Greek *phulon*, meaning race, and *gen*, meaning produced by.

Phylum A major division of the animal kingdom. Plural: phyla.

Plankton Animals and plants that float or swim weakly in a body of water; from the Greek *planktos*, meaning wandering or roaming.

Planktotrophic larva A larva that feeds on other planktonic material rather than relying on stored food.

Polychaete worms Segmented worms (Phylum Annelida) characterized by numerous bristles or flaps along the sides of the body.

Postembryonic development Changes that occur after the embryo has broken through the egg membrane and become a larva.

Prismatic Flat-sided, like a prism.

Proximal Closest to the centre or central axis.

Pseudopaxillae Aboral calcareous plates similar to true paxillae but their spines or spinelets are fixed, with no muscular control. Characteristic of the Family Solasteridae and others (figure 5).

Pyloric caeca Paired glandular organs in the arms of a sea star that secrete mucus and digestive enzymes, and serve as a food storage organs (figures 7 and 8). Singular: pyloric caecum.

Pyloric ducts Tubes that join the pyloric caeca to the pyloric stomach (figure 7).

Pyloric stomach The most aboral of a sea star's two stomachs, it is connected to the pyloric caeca by the pyloric ducts (figure 7).

Radial nerve cord A major nerve that runs along the ambulacral groove of each arm (figure 8).

Radial symmetry A repeating arrangement of similar parts around a central axis.

Radial water canal A branch of the water-vascular system that runs the length of each arm (figure 8).

Recruitment The addition of new individuals to a population by successful spawning and the settlement of juveniles.

Reticulate Having the structure or appearance of a netlike pattern of ridges or fibres.

Reticulum A network or weblike structure.
Ring canal Part of the water-vascular system that surrounds the mouth (figure 6).
R value The distance from the centre of the disc to the tip of an arm (figure 3).
r value The distance from the centre of the disc to the edge of the disc between two arms (figure 3).
Scale worm A family of polychaete worms characterized by a series of overlapping plates on the back.
Sea squirts Filter-feeding animals that are primitive members of the Phylum Chordata. Also called ascidians or tunicates; they can be solitary or colonial.
Species A set of similar individuals capable of breeding successfully with one another in nature, and reproductively isolated from other sets.
Spinelet A short, stubby spine.
Spiracles Small contractile pores in the supradorsal membrane of the Family Pterasteridae (figure 80).
Stone canal A calcareous tube that connects the madreporite to the ring canal of the water-vascular system (figure 6).
Subcapitate Not quite capitate.
Suboral spines Spines situated on the oral surface of the mouth plates rather than on the edge (figure 28).
Subspecies Part of a species population that is physically distinct and confined to a geographic area within the range of the species.
Substrate The surface or material on which an organism grows or lives. Plural: substrates.
Subtidal Below the level of the lowest tide.
Superambulacral plate An internal calcareous plate lying across the inner junction of ambulacral and adambulacral plates in some sea stars (figure 27); not visible without dissection.
Supradorsal membrane A thin covering supported by the tips of the pseudopaxillar spines in Family Pterasteridae (figure 80).
Superomarginals A series of plates that form the upper lateral side or edge of the arm (figure 27); also called supra, superior or upper marginals.
Suture A joint between two calcareous plates, as in the median suture between two adjoining mouth plates (figure 28).
Synonym A different scientific name for the same species. The earliest published name takes precedence.
Taxonomic Relating to taxonomy, the study of the classification of plants and animals according to natural relationships; from the Greek *tasso*, meaning arrange or classify.
Tiedemann's Bodies Blind sacs connected to the ring canal of the water-vascular system that take part in purifying incoming sea water and producing coelomic fluid.

Transverse series A row of spines running at right angles to the ambulacral furrow or a central axis (figure 29).

Trinomen A three-part scientific name consisting of the genus, species and subspecies; e.g., *Henricia leviuscula annectens*.

Tube feet Extendible tubular appendages, also called podia, located in the ambulacral furrow. Tube feet may be pointed or bear a sucker at the tip (figure 8).

Tunicate A filter-feeding animal that is a primitive member of the Phylum Chordata. Also called ascidian or sea squirt; it can be solitary or colonial.

Tusk shells Tusk-shaped molluscs that live in soft substrates with only the narrow tip projecting from the sand (Class Scaphopoda).

Type locality The place where the holotype was collected.

Water-vascular system An enclosed network of tubes and sacs that operates the tube feet by hydraulic pressure (figure 6).

REFERENCES

Ahearn, G.A. 1990. Proline transport by brush-border membrane vesicles of a starfish pyloric caeca. *Journal of Experimental Biology* 150:453-59.

Ahearn, G.A., and R.D. Behnke. 1991. L-proline transport systems of starfish pyloric caeca. *Journal of Experimental Biology* 158:477-93.

Alexander, M.E., and M.H. Dresden. 1980. Collagenolytic enzymes from the starfish *Pycnopodia helianthoides. Comparative Biochemistry and Physiology Series B* 67:505-9.

Alton, M.S. 1966. Bathymetric distribution of sea stars (Asteroidea) off the northern Oregon coast. *Journal of the Fisheries Research Board of Canada* 23:1673-1714.

Anderson, J.M. 1959. Studies on the cardiac stomach of a starfish, *Patiria miniata* (Brandt). *Biological Bulletin* (Woods Hole, Mass.) 117:185-201.

———. 1960. Histological studies on the digestive system of a starfish *Henricia*, with notes on Tiedemann's pouches in starfishes. *Biological Bulletin* (Woods Hole, Mass.) 119:371-98.

Anderson, R.C., and R.L. Shimek. 1993. A note on the feeding habits of some uncommon sea stars. *Zoo Biology* 12:499-503.

Annett, C., and R. Pierotti. 1984. Foraging behaviour and prey selection of the leather seastar *Dermasterias imbricata. Marine Ecology Progress Series* 14:197-206.

Asotra, S., P.V. Mladenov and R.D. Burke. 1988. Polyamines and cell proliferation in the sea star *Pycnopodia helianthoides. Comparative Biochemistry and Physiology Series B* 90:885-90.

Austin, W.C. 1985. *An annotated checklist of marine invertebrates in the cold temperate northeast Pacific.* Cowichan Bay, B.C.: Khoyatan Marine Laboratory.

Baker, A.N., F.W.E. Rowe and H.E.S. Clark. 1986. A new class of Echinodermata from New Zealand. *Nature* (London) 321:862-64.

Bakus, G., J. 1974. Toxicity in holothurians: A geographical pattern. *Biotropica* 6:229-36.

Banfield, D.K., J.D.G. Boom, B.M. Honda and M.J. Smith. 1988. H3 Histone RNA in eggs and embryos of the sea star *Pisaster ochraceus*. *Biochemistry and Cell Biology* 66:1040-44.

Barber, M.L. 1979. Changes in enzyme activities and lipid content of echinoderm egg membranes, at maturation and fertilization. *American Zoologist* 19:821-37.

Barr, L., and N. Barr. 1983. *Under Alaskan seas*. Anchorage: Alaska Northwest Publishing.

Bavendam, F. 1985. Sea stars deploy a bag of tricks in marine wars. *Smithsonian* 16:104-9.

Bell, M.V., and J.R. Sargent. 1985. Fatty acid analyses of phosphoglycerides from tissues of the clam *Chlamys islandica* (Muller) and the starfish *Ctenodiscus crispatus* (Retzius) from Balsfjorden, northern Norway. *Journal of Experimental Marine Biology and Ecology* 87:31-40.

Bigger, C.H., and W.H. Hildemann. 1982. Leukocytes of the sea star, *Dermasterias imbricata*. Pp. 355. In *Echinodermata: Proceedings of the Third International Echinoderm Conference, Tampa Bay*, edited by J.M. Lawrence. Rotterdam: A.A. Balkema.

Bingham, B.L., and L.F. Braithwaite. 1986. Defence adaptations of the Dendrochirote Holothurian *Psolus chitonoides* Clark. *Journal of Experimental Marine Biology and Ecology* 98:311-22.

Birkeland, C., F.S. Chia and R.R. Strathmann. 1971. Development, substratum selection, delay of metamorphosis and growth in the seastar, *Mediaster aequalis* Stimpson. *Biological Bulletin* (Woods Hole, Mass.) 141:99-108.

Birkeland, C., P.K. Dayton and N.A. Engstrom. 1982. A stable system of predation on a holothurian by four asteroids and their top predator. *Australian Museum Memoir* 16:175-89.

Blake, D.B. 1987. A classification and phylogeny of post-Palaeozoic sea stars (Asteroidea: Echinodermata). *Journal of Natural History* 21:481-528.

Boolootian, R.A. 1966. *Physiology of Echinodermata*. New York: Interscience Publishers.

Boom, J.D.G., and M.J. Smith. 1989. Molecular analyses of gene expression during sea star spermatogenesis. *Journal of Experimental Zoology* 250:312-20.

Brandenburger, J.L., and R.M. Eakin. 1980. Cytochemical localization of Acid phosphatase in ocelli of the sea star *Patiria miniata* during recycling of photoreceptoral membranes. *Journal of Experimental Zoology* 214:127-40.

Bruno, I., L. Minale and R. Riccio. 1989. Starfish saponins, 38. Steroidal glycosides from the starfish *Pycnopodia helianthoides*. *Journal of Natural Products (Lloydia)* 52:1022-26.

———. 1990. Starfish saponins, part 43. Structures of two new sulfated steroidal fucofuranosides (Imbricatosides A and B) and six new polyhydroxysteroids from the starfish *Dermasterias imbricata*. *Journal of Natural Products (Lloydia)* 53:366-74.

Bruno, I., M.V.D. d'Auria, M. Iorizzi, L. Minale and R. Riccio. 1992. Marine eicosanoids: occurrence of 8,11,12-trihydroxylated eicosanoic acids in starfishes. *Experientia* 48:114-15.

Burke, R.D. 1983. The structure of the larval nervous system of *Pisaster ochraceus* (Echinodermata: Asteroidea). *Journal of Morphology* 178:23-25.

Burke, R.D., and R.F. Watkins. 1991. Stimulation of starfish coelomocytes by interleukin-1. *Biochemical and Biophysical Research Communications* 180:579-84.

Burnell, D.J., J.W. ApSimon and M.W. Gilgan. 1986. Occurrence of saponins giving rise to asterone and asterogenol in various species of starfish. *Comparative Biochemistry and Physiology Series B* 85:389-91.

Byrne, M., A. Cerra, T. Nishigaki and M. Hoshi. 1998. Male infertility in *Asterias amurensis*: A new phenomenon resulting from introduction of the parasitic ciliate *Orchitophrya stellarum* to Japan. Pp. 203-7. In *Echinoderms: Proceedings of the Ninth International Echinoderm Conference, San Francisco, California, U.S.A. 5-9 August 1996*, edited by R. Mooi and M. Telford. Rotterdam: A. A. Balkema.

Caine, G.D., and R.D. Burke. 1985. Immunohistochemical localization of gonad stimulating substance in the seastar *Pycnopodia helianthoides*. Pp. 495-98. In *Echinodermata: Proceedings of the Fifth International Echinoderm Conference, Galway/24-29, September 1984*, edited by B. Keegan F and D.S. O'Connor. Rotterdam: A.A. Balkema.

Cameron, R.A., and N.D. Holland. 1983. Electron microscopy of extra-cellular materials during the development of a sea star, *Patiria miniata* (Echinodermata: Asteroidea). *Cell and Tissue Research* 234:193-200.

Campbell, S.S., and B.J. Crawford. 1991. Ultrastructural study of the hyaline layer of the starfish embryo, *Pisaster ochraceus*. *Anatomical Record* 231:125-35.

Carey, A.J. 1972. Food sources of sublittoral, bathyal and abyssal asteroids in the northeast Pacific Ocean. *Ophelia* 10:35-47.

Carlson, H.R., and C.A. Pfister. 1999. A seventeen-year study of the rose star *Crossaster papposus* population in a coastal bay in southeast Alaska. *Marine Biology* (Berlin) 133:223-30.

Carson, S.F. 1988. Development and metamorphosis of three species of the asteroid *Solaster* from the northeast Pacific. Pp. 791. In *Echinoderm Biology*, edited by R.D. Burke, P.V. Mladenov, P. Lambert and R.L. Parsley. Rotterdam/Brookfield: A. A. Balkema.

Cavey, M.J., and C.A. Sokoluk. 1987. Heteromorphic discocilia of the phanerozonian starfish *Luidia foliolata*: artifacts of chemical fixation. *Transactions of the American Microscopical Society* 106:1-9.

Cavey, M.J., and R.L. Wood. 1981. Specializations for excitation-contraction coupling in the podial retractor cells of the starfish, *Stylasterias forreri*. *Cell and Tissue Research* 218:475-85.

———. 1988. Sarcolemmal morphometry in the podial retractor cells of non-burrowing and burrowing starfishes. Pp. 707-12. In *Echinoderm*

Biology: Proceedings of the sixth international echinoderm conference, Victoria/23-28 August, 1987, edited by R.D. Burke, P.V. Mladenov, P. Lambert and R.L. Parsley. Rotterdam: A.A. Balkema.

———. 1991. Organization of the adluminal and retractor cells in the coelomic lining from the tube foot of a phanerozonian starfish, *Luidia foliolata*. *Canadian Journal of Zoology* 69:911-23.

Chaffee, C., and R.B. Spies. 1982. The effects of used ferrochrome lignosulphonate drilling muds from a Santa Barbara channel oil well on the development of starfish embryos. *Marine Environmental Research* 7:265-77.

Chia, F.S. 1966a. Brooding behaviour of a six-rayed starfish *Leptasterias hexactis*. *Biological Bulletin* (Woods Hole, Mass.) 130:304-15.

———. 1966b. Development of a deep-sea cushion star, *Pteraster tesselatus*. *Proceedings of the California Academy of Sciences* 34:505-10.

———. 1966c. Systematics of the six rayed sea star, *Leptasterias* in the vicinity of the San Juan Island, Washington. *Systematic Zoology* 15:300-306.

———. 1968a. The embryology of a brooding starfish, *Leptasterias hexactis* (Stimpson). *Acta Zoologica (Stockholm)* 49:321-64.

———. 1968b. Some observations on the development and cyclic changes of the oöcytes in a brooding starfish, *Leptasterias hexactis*. *Journal of Zoology* 154:453-61.

Chia, F.S., and H. Amerongen. 1975. On the prey-catching pedicellariae of a starfish, *Stylasterias forreri* (de Loriol). *Canadian Journal of Zoology* 53:748-55.

Chia, F.S., and R. Koss. 1994. Asteroidea. Pp. 169-246. In *Microscopic anatomy of invertebrates*, vol. 14, edited by F.W. Harrison and F.S. Chia. Toronto: Wiley-Liss.

Chia, F.S., and C.W. Walker. 1991. Echinodermata: Asteroidea. Pp. 301-53. In *Reproduction of marine invertebrates - Echinoderms and Lophophorates*, vol. 6, edited by A.C. Giese, J.S. Pearse and V.B. Pearse. Pacific Grove, California: The Boxwood Press.

Chiba, K., R.T. Kado and L.A. Jaffe. 1990. Development of calcium release mechanisms during starfish oocyte maturation. *Developmental Biology* 140:300-306.

Christen, R. 1985. Isolation of acrosomal vesicles and their surrounding membranes from starfish sperm. *Development, Growth and Differentiation* 27:529-38.

Christensen, A.M. 1957. The feeding behaviour of the seastar *Evasterias troschelii* (Stimpson). *Limnology and Oceanography* 2:180-97.

Clark, A.M. 1981. Notes on Atlantic and other Asteroidea. 1. Family Benthopectinidae. *Bulletin of the British Museum of Natural History (Zoology)* 41:91-135.

———. 1983. Notes on Atlantic and other Asteroidea. 3. The families Ganeriidae and Asterinidae, with description of a new asterinid genus. *Bulletin of the British Museum of Natural History (Zoology)* 45:359-80.

Clark, A.M. 1984. Notes on the Atlantic and other Asteroidea. 4. Families Poraniidae and Asteropseidae. *Bulletin of the British Museum of Natural History (Zoology)* 47:19-51.

———. 1989. An index of names of recent Asteroidea - Part 1: Paxillosida and Notomyotida. Pp. 225-347. In *Echinoderm Studies*, vol. 3, edited by M. Jangoux and J.M. Lawrence. Rotterdam: A. A. Balkema.

———. 1993. An index of names of recent Asteroidea - Part 2: Valvatida. Pp. 187-366. In *Echinoderm Studies*, vol. 4, edited by M. Jangoux and J.M. Lawrence. Rotterdam/ Brookfield: A. A. Balkema.

———. 1996. An index of names of recent Asteroidea - Part 3: Velatida and Spinulosida. Pp. 183-250. In *Echinoderm Studies*, vol. 5, edited by M. Jangoux and J.M. Lawrence. Rotterdam/ Brookfield: A. A. Balkema.

Clark, A.M., and M.E. Downey. 1992. *Starfishes of the Atlantic*. London: Chapman and Hall.

Cloud, J.G. 1983. Hydrostatic pressure inhibition of hormone-induced resumption of meiotic maturation in isolated oocytes. *Journal of Experimental Zoology* 227:265-70.

———. 1984. Pressure inhibition of starfish oocyte maturation: Temporal aspects. *Journal of Experimental Zoology* 230:17-21.

Collier, J.R. 1983. The molecular weight of ribosomal ribonucleic acids among the Protostomia. *Biological Bulletin* (Woods Hole, Mass.) 164:428-32.

Cool, D., D. Banfield, B.M. Honda and M.J. Smith. 1988. Histone genes in three sea star species: cluster arrangement, transcriptional polarity, and analyses of the flanking regions of H3 and H4 genes. *Journal of Molecular Evolution* 27:36-44.

Cowden, C., C.M. Young and F.S. Chia. 1984. Differential predation on marine invertebrate larvae by two benthic predators. *Marine Ecology Progress Series* 14:145-49.

Crawford, B.J. 1988. Regional ultrastructural differences in basal laminae isolated from the starfish, *Pisaster ochraceus*. *Development, Growth and Differentiation* 30:661-72.

———. 1989. Ultrastructure of the basal lamina and its relationship to extracellular matrix of embryos of the starfish *Pisaster ochraceus* as revealed by anionic dyes. *Journal of Morphology* 199:349-61.

———. 1990. Changes in the arrangement of the extracellular matrix, larval shape, and mesenchyme cell migration during asteroid larval development. *Journal of Morphology* 260:147-61.

Crawford, B., and M. Abed. 1983. The role of the basal lamina in mouth formation in the embryo of the starfish, *Pisaster ochraceus*. *Journal of Morphology* 176:235-46.

Crawford, B.J., and F.S. Chia. 1978. Coelomic pouch formation in the starfish *Pisaster ochraceus* (Echinodermata: Asteroidea). *Journal of Morphology* 157:99-120.

———. 1980. Induction and development of exogastrulae in the Starfish *Pisaster ochraceus*. *Journal of Invertebrate Reproduction* 2:37-46.

———. 1982. Genesis and movement of mesenchyme cells in embryos of the starfish *Pisaster ochraceus*. Pp. 505-11. In *Echinoderms: Proceedings of the Third International Echinoderm Conference, Tampa Bay*, edited by J.M. Lawrence. Rotterdam: A.A. Balkema.

Crawford, T.J., and B.J. Crawford. 1992. Characterization and localization of large sulfated glycoproteins in the extracellular matrix of the developing asteroid *Pisaster ochraceus. Biochemistry and Cell Biology* 70:91-98.

Dalby, J.E., Jr., J.K. Elliott and D.M. Ross. 1988. The swim response of the actinian *Stomphia didemon* to certain asteroids: distribution and phylogenetic implications. *Canadian Journal of Zoology* 66:2484-91.

d'Auria, M.V., L. Minale, R. Riccio and E. Uriarte. 1988. Marine eicosanoids: occurrence of 8-(R)-Hete in the starfish *Patiria miniata. Experientia* 44:719-20.

d'Auria, M.V., M. Iorizzi, L. Minale and R. Riccio. 1990a. Starfish saponins, part 40. Structures of two new "asterosaponins" from the starfish *Patiria miniata*: patirioside. *Journal of the Chemical Society. Perkin Transactions* 1990:1019-23.

d'Auria, M.V., M. Iorizzi, L. Minale, R. Riccio and E. Uriarte. 1990b. Starfish saponins, part 41. Structure of two new steroidal glycoside sulfates (miniatosides A and B) and two new polyhydroxysteroids from the starfish *Patiria miniata. Journal of Natural Products (Lloydia)* 53:94-101.

David, B., and R. Mooi. 1998. Major events in the evolution of echinoderms viewed by the light of embryology. Pp. 923. In *Echinoderms: San Francisco. Proceedings of the Ninth International Echinoderm Conference, San Francisco, California, USA. 5-9 August 1996*, edited by R. Mooi and M. Telford. Rotterdam: A.A. Balkema.

Davis, K.K. 1985. DNA synthesis and the annual spermatogenic cycle in individuals of the sea star, *Patiria miniata. Biological Bulletin* (Woods Hole, Mass.) 169:313-27.

Davis, P.H., T.W. Schultz and R.B. Spies. 1981. Toxicity of Santa Barbara Seep oil to starfish embryos. Part 2: the growth bioassay. *Marine Environmental Research* 5:287-94.

Day, R.W., and R.W. Osman. 1981. Predation by *Patiria miniata* (Asteroidea) on Bryozoans: prey diversity may depend on the mechanism of succession. *Oecologia* 51:300-309.

Daya-Makin, M., S.L. Pelech, A. Levitzki and A.T. Hudson. 1991. Erbstatin and tyrphostins block protein-serine kinase activation and meiotic maturation of sea star oocytes. *Biochimica et Biophysica Acta* 1093:87-94.

Dembetskii, V.M. 1988. Phospholipid and fatty acid composition in echinoderms. 2. Class Asteroidea. *Khimiya Prirodnykh Soedinenii* 1988:353-57,471.

DeSantis, M., and J.G. Cloud. 1984. Electrical stimulation of the starfish's radial nerve in vitro induces the release of a gonadotrophin. *Journal of Experimental Zoology* 231:423-27.

———. 1988. Ovarian sensitivity to radial nerve factor in starfish. Pp. 551-55. In *Echinoderm Biology: Proceedings of the sixth international echinoderm conference, Victoria/23-28 August, 1987*, edited by R.D. Burke, P.V. Mladenov, P. Lambert and R.L. Parsley. Rotterdam: A.A. Balkema.

Doering, P.H., and D.W. Phillips. 1983. Maintenance of the shore-level size gradient in the marine snail *Tegula funebralis* (A. Adams): importance of behavioural responses to light and seastar predators. *Journal of Experimental Marine Biology and Ecology* 67:159-73.

Duben, M.W., von., and J. Koren. 1846. Ofversigt af Skandinaviens Echinodermer. *Sv. Vetensk. Akad. Handl.* for 1844:229-328.

Duggins, D.O. 1983. Starfish predation and the creation of mosaic patterns in a kelp-dominated community. *Ecology* 64:1610-19.

D'Yakonov, A.M. 1950. *Keys to the Fauna of the USSR. No.34. Sea Stars (Asteroids) of the USSR Seas.* USSR / Jerusalem: Zoological Institute of the Academy of Sciences of the USSR (translated 1968 by Israel Program for Scientific Translations, Jerusalem).

Eakin, R.M., and J.L. Brandenburger. 1979. Effects of light on ocelli of seastars. *Zoomorphology* (Berlin) 92:191-200.

Edwards, D.C. 1969. Predators on *Olivella biplicata*, including a species-specific predator avoidance response. *The Veliger* 11:326-33.

Elliott, J., J. Dalby Jr, R. Cohen and D.M. Ross. 1985. Behavioural interactions between the actinian *Tealia piscivora* (Anthozoa: Actinaria) and the asteroid *Dermasterias imbricata*. *Canadian Journal of Zoology* 63:1921-29.

Elliott, J.K., D.M. Ross, C. Pathirana, S. Miao, R.J. Anderson, P. Singer, W.C.M.C. Kokke and W.A. Ayer. 1989. Induction of swimming in *Stomphia* (Anthozoa: Actinaria) by imbricatine, a metabolite of the asteroid *Dermasterias imbricata*. *Biological Bulletin* (Woods Hole, Mass.) 176:73-78.

Engstrom, N.A. 1988. The role of a predator's prey preference in maintaining natural monocultures of a subtidal holothurian: sweet cucumbers, sour cucumbers, a cucumber connoisseur, and pickle patches. Pp. 445-48. In *Echinoderm Biology*, edited by R.D. Burke, P.V. Mladenov, P. Lambert and R.L. Parsley. Rotterdam: A.A. Balkema.

Estell, D.A., and M. Laskowski Jr. 1980. *Dermasterias imbricata* trypsin 1: an enzyme which rapidly hydrolyzes the reactive-site peptide bonds of protein trypsin inhibitors. *Biochemistry* 19:124-31.

Estell, D.A., K.A. Wilson and M. Laskowski Jr. 1980. Thermodynamics and kinetics of the hydrolysis of the reactive-site peptide bond in pancreatic trypsin inhibitor (Kunitz) by *Dermasterias imbricata* trypsin 1. *Biochemistry* 19:131-37.

Falk-Petersen, I.-B. 1982. Reproductive and biochemical studies of the asteroid *Ctenodiscus crispatus* (Retzius). *Sarsia* 67:123-30.

Falk-Peterson, I.-B., and J.R. Sargent. 1982. Reproduction of Asteroids from Balsfjorden, northern Norway: analyses of lipids in the gonads of *Ctenodiscus crispatus, Asterias lincki*, and *Pteraster militaris. Marine*

Biology (Berlin) 69:291-98.
Farmanfarmaian, A., A.C. Giese, R.A. Boolootian and J. Bennett. 1958. Annual reproductive cycles in four species of west coast starfishes. *Journal of Experimental Zoology* 138:355-67.
Farrand, A.L., and D.C. Williams. 1988. Isolation, purification and partial characterization of four digestive proteases from the purple sea star *Pisaster ochraceus*. *Marine Biology* (Berlin) 97:231-36.
Fawcett, M.H. 1984. Local and latitudinal variation in predation on an herbivorous marine snail. *Ecology* 65:1214-30.
Feder, H.M. 1955. On the methods used by the starfish *Pisaster ochraceus* in opening three types of bivalve mollusks. *Ecology* 36:764-67.
———. 1959. The food of the starfish *Pisaster ochraceus*, along the California Coast. *Ecology* 40:721-24.
———. 1963. Gastropod defensive responses and their effectiveness in reducing predation by starfishes. *Ecology* 44:505-12.
———. 1970. Growth and predation by the ochre sea star, *Pisaster ochraceus* (Brandt) in Monterey Bay, California. *Ophelia* 8:161-85.
Ferguson, J.C. 1967. An autoradiographic study of the utilization of free exogenous amino acids by starfishes. *Biological Bulletin* (Woods Hole, Mass.) 133:317-29.
———. 1990. Seawater inflow through the madreporite and internal body regions of a starfish (*Leptasterias hexactis*) as demonstrated with fluorescent microbeads. *Journal of Experimental Zoology* 255:262-71.
———. 1992. The function of the madreporite in body fluid volume maintenance by an intertidal starfish, *Pisaster ochraceus*. *Biological Bulletin* (Woods Hole, Mass.) 183:482-89.
———. 1994. Madreporite inflow of seawater to maintain body fluids in five species of starfish. Pp. 285-89. In *Echinoderms through time. Proceedings of the Eighth International Echinoderm Conference, Dijon, France, 6-10 September 1993*, edited by B. David, A. Guille, J.P. Feral and M. Roux. Rotterdam and Brookfield: A. A. Balkema.
Ferguson, J.C., and C.W. Walker. 1991. Cytology and function of the madreporite systems of the starfish *Henricia sanguinolenta* and *Asterias vulgaris*. *Journal of Morphology* 210:1-11.
———. 1993. Adhesion seams in the Tiedemann's diverticula of the starfish *Henricia sanguinolenta*. *Transactions of the American Microscopical Society* 112:158-67.
Ferrand, J.G. 1982. The activities of hydrolases in ova and sperm of starfishes. Pp. 485-88. In *Echinodermata: Proceedings of the Third International Echinoderm Conference, Tampa Bay*, edited by J.M. Lawrence. Rotterdam: A.A. Balkema.
Fisher, W.K. 1908. Necessary changes in the nomenclature of starfishes. *Smithsonian Miscellaneous Collections* 52:87-93.
———. 1910. New starfishes from the north Pacific. 1. Phanerozonia. *Zoologische Anzeiger* 35:546-53.

Fisher, W.K. 1911. *Asteroidea of the North Pacific and adjacent waters. Part 1. Phanerozonia and Spinulosa*. Washington, D.C.: Smithsonian Institution, U.S. National Museum.

———. 1923. A preliminary synopsis of the Asteriidae, a Family of sea stars. *Annals and Magazine of Natural History, Series 9* 10 and 12:247-98, 596-607.

———. 1928. *Asteroidea of the North Pacific and adjacent waters. Part 2. Forcipulata (part)*. Washington, D.C.: Smithsonian Institution, U.S. National Museum.

———. 1930. *Asteroidea of the North Pacific and adjacent waters. Part 3. Forcipulata (concluded)*. Washington, D.C.: Smithsonian Institution, U.S. National Museum.

Fishlyn, D.A., and D.W. Phillips. 1980. Chemical camouflaging and behavioural defenses against a predatory sea star by three species of gastropods from the surfgrass *Phyllospadix* community. *Biological Bulletin* (Woods Hole, Mass.) 158:34-48.

Flowers, J.M. 1999. Discordant patterns of genetic and morphological variation and their implications for the taxonomy of *Leptasterias* subgenus *Hexasterias* of the North Pacific. M.Sc. thesis. Louisiana State University, Baton Rouge.

Foltz, D.W. 1997. Hybridization frequency is negatively correlated with divergence time of mitochondrial DNA haplotypes in a sea star (Leptasterias spp.) species complex. *Evolution* 51:283-88.

———. 1998. Distribution of intertidal *Leptasterias* spp. along the Pacific North American coast: A synthesis of allozymic and mtDNA data. Pp. 923. In *Echinoderms: Proceedings of the Ninth International Echinoderm Conference, San Francisco, California*, edited by R. Mooi and M. Telford. Rotterdam/ Brookfield: A. A. Balkema.

Foltz, D.W., and W.B. Stickle Jr. 1994. Genetic structure of four species in the *Leptasterias hexactis* complex along the Pacific coast of North America. Pp. 291-96. In *Echinoderms through time. Proceedings of the Eighth International Echinoderm Conference, Dijon, France, 6-10 September 1993*, edited by B. David, A. Guille, J.P. Feral and M. Roux. Rotterdam: A.A. Balkema.

Foltz, D.W., J.P. Breaux, E.L. Campagnaro, S.W. Herke, A.E. Himel, A.W. Hrincevich, J.W. Tamplin and W.B. Stickle. 1996a. Limited morphological differences between genetically identified cryptic species within the *Leptasterias* species complex (Echinodermata: Asteroidea). *Canadian Journal of Zoology* 74:1275-83.

Foltz, D.W., W.B. Stickle, E.L. Campagnaro and A.E. Himel. 1996b. Mitochondrial DNA polymorphisms reveal additional genetic hereogeneity within the *Leptasterias hexactis* (Echinodermata: Asteroidea) species complex. *Marine Biology* (Berlin) 125:569-78.

Forbes, E. 1839. On the Asteriadae of the Irish Sea. *Memoirs of the Wernerian Society*, Edinburgh 8:114-29.

Francour, P. 1997. Predation on holothurians: a literature review. *Invertebrate Biology* 116:52-60.
Fraser, A., J. Gomez, E.B. Hartwick and M.J. Smith. 1981. Observations on the reproduction and development of *Pisaster ochraceus* (Brandt). *Canadian Journal of Zoology* 59:1700-1707.
Garnas, R.L., and D.G. Crosby. 1979. Comparative metabolism of parathion by intertidal invertebrates. Pp. 291-305. In *Marine Pollution: functional responses*, edited by W.B. Vernberg, A. Calabrese, F.P. Thurberg and F.J. Vernberg. New York, San Fransisco, and London: Academic press.
Gemmill, J.F. 1912. The development of the starfish *Solaster endeca* (Forbs). *Transactions of the Zoological Society of London* 20:1-58.
———. 1920. The development of the starfish *Crossaster papposus*, Muller and Troschel. *Quarterly Journal of Microscopical Science* 64:155-89.
George, S.B. 1994. The *Leptasterias* (Echinodermata: Asteroidea) species complex: variation in reproductive investment. *Marine Ecology Progress Series* 109:95-98.
Gotshall, D.W. 1994. *Guide to marine invertebrates - Alaska to Baja California.* Monterey, California: Sea Challengers.
Grainger, E.H. 1966. Sea Stars (Echinodermata: Asteroidea) of Arctic North America. *Fisheries Research Board of Canada Bulletin* No. 152:1-70.
Greer, D.L. 1962. Studies on the embryology of *Pycnopodia helianthoides* (Brandt) Stimpson. *Pacific Science* 16:280-85.
Grieg, J.A. 1921. Echinodermata from the Michael Sars North Atlantic Deep Sea Expedition 1910. *Report of the "Michael Sars" North Atlantic Deep-Sea Expedition 1910* 3:44 pp.
Grygier, M.J. 1982. *Dendrogaster* (Crustacea: Ascothoracida) from California: sea star parasites collected by the "Albatross". *Proceedings of the California Academy of Sciences* 42:443-54.
Haderlie, E.C. 1980. Sea star predation on rock-boring bivalves. *The Veliger* 22:400.
Hagiwara, S. 1979. Effects of internal potassium and sodium on the anomalous rectification of the starfish egg as examined by internal perfusion. *Journal of Physiology*(London) 292:251-65.
Halberg, F., F. Halberg, R.B. Sothern, J.S. Pearse, V.B. Pearse, K. Shankaraiah and A.C. Giese. 1987. Consistent synchronization and circaseptennian (about seven yearly) modulation of circannual gonadal index rhythm of three marine invertebrates. *Progress in Clinical and Biological Research* 227:225-38.
Halpern, J.A. 1972. Pseudarchasterinae (Echinodermata: Asteroidea) of the Atlantic. *Proceedings of the Biological Society of Washington* 85:359-84.
Hart, M.W. 1991. Particle captures and the method of suspension feeding by echinoderm larvae. *Biological Bulletin* (Woods Hole, Mass.) 180:12-27.
Harvey, R., J.D. Gage, D.S.M. Billet, A.M. Clark and G.L. Paterson. 1988. Echinoderms of the Rockall Trough and Adjacent areas. 3. Additional

records. *Bulletin of the British Museum of Natural History (Zoology)* 54:153-98.
Heath, H. 1917. The early development of a starfish *Pateria*[sic] (*Asterina*) *miniata*. *Journal of Morphology* 29:461-68 (1pl.).
Highsmith, R.C. 1985. Floating and algal rafting as potential dispersal mechanisms in brooding invertebrates. *Marine Ecology Progress Series* 25:169-79.
Himmelman, J.H. 1991. Diving observations of subtidal communities in the northern Gulf of St. Lawrence. *Canadian special publications of fisheries and aquatic sciences* 113:319-32.
Himmelman, J.H., and C. Dutil. 1991. Distribution, population structure and feeding of subtidal seastars in the northern Gulf of St. Lawrence. *Marine Ecology Progress Series* 76:61-72.
Hoffman, D.L. 1980. Defensive responses of marine gastropods (Prosobranchia, Trochidae) to certain predatory sea stars and the Dire whelk, *Searlesia dira* (Reeve). *Pacific Science* 34:233-43.
Holland, N.D. 1980. Electron microscopic study of the cortical reaction in eggs of the starfish (*Patiria miniata*). *Cell and Tissue Research* 205:67-76.
Holme, N.A. 1966. The bottom fauna of the English Channel (Part 2). *Journal of the Marine Biological Association of the United Kingdom* 46:401-93.
Holzman, T.F., S.F. Russo and D.C. Williams. 1985. Effects of feeding and starvation on proteolytic and tryptic activities in pyloric caecal tissues and duct fluids of the sea star *Pisaster ochraceus*. *Marine Biology* (Berlin) 90:55-59.
Hopkins, T.S., and G.F. Crozier. 1966. Observations on the asteroid echinoderm fauna occurring in the shallow water of Southern California (intertidal to 60 metres). *Bulletin of the Southern California Academy of Sciences* 65:129-45.
Horowitz, R.A., D.A. Agard, J.W. Sedat and C.L. Woodcock. 1994. The three-dimensional architecture of chromatin in situ: electron tomography reveals fibers composed of a continuously variable zig-zag nucleosomal ribbon. *Journal of Cell Biology* 125:1-10.
Howell, A.M., D. Cool, J. Hewitt, B. Ydenberg, M.J. Smith and B.M. Honda. 1987. Organization and unusual expression of histone genes in the seastar *Pisaster ochraceus*. *Journal of Molecular Evolution* 25:29-36.
Hrincevich, A.W., and D.W. Foltz. 1996. Mitochondrial DNA sequence variation in a sea star (*Leptasterias* spp.) species complex. *Molecular Phylogenetics and Evolution* 6:408-15.
Huvard, A.L., and N.D. Holland. 1986. Pinocytosis of ferritin from the gut lumen in larvae of a seastar (*Patiria miniata*) and a sea urchin (*Lytechinus pictus*). *Development, Growth and Differentiation* 28:43-57.
Hyman, L.H. 1955. *The Invertebrates: Echinodermata - The Coelomate Bilateria Vol. IV*. Toronto, Ont.: McGraw-Hill Book Co.
Jacobs, H.T., S. Asakawa, T. Araki, K.-I. Miura, M.J. Smith and K. Watanabe. 1989. Conserved tRNA gene cluster in starfish mitochondrial DNA. *Current Genetics* 15:193-206.

Jangoux, M. 1982. Food and feeding mechanisms: Asteroidea. Pp. 117-59. In *Echinoderm nutrition*, edited by M. Jangoux and J.M. Lawrence. Rotterdam: A. A. Balkema.
Janies, D.A., and L.R. McEdward. 1993. Highly derived coelomic and water-vascular morphogenesis in a starfish with pelagic direct development. *Biological Bulletin* (Woods Hole, Mass.) 185:56-76.
———. 1994. Heterotopy, pelagic direct development, and new body plans in velatid asteroids. Pp. 319-24. In *Echinoderms through time. Proceedings of the Eighth International Echinoderm Conference, Dijon, France, 6-10 September 1993*, edited by B. David, A. Guille, J.P. Feral and M. Roux. Rotterdam: A.A. Balkema.
Johansen, K., and J.A. Petersen. 1971. Gas exchange and active ventilation in a starfish *Pteraster tesselatus*. *Z. Vergl. Physiol.* 71:365-81.
Johnson, C.H., and D. Epel. 1982. Starfish oocyte maturation and fertilization: intracellular pH is not involved in activation. *Developmental Biology* 92:461-69.
Jonas-Davies, J., and J. Liston. 1985. The occurrence of PSP toxins in intertidal organisms. Pp. 467-72. In *Toxic dinoflagellates*, edited by D.M. Anderson, A.W. White and D.G. Baden. New York, Amsterdam, and Oxford: Elsevier Science Publishing Co.
Kaneshiro, E.S., and R.D. Karp. 1980. The ultrastructure of coelomocytes of the sea star *Dermasterias imbricata*. *Biological Bulletin* (Woods Hole, Mass.) 159:295-310.
Kasyanov, V.L., G.A. Kryuchkova, V.A. Kulikova and L.A. Medvedeva. 1998. *Larvae of marine bivalves and echinoderms*. Washington, D.C.: Smithsonian Institution Libraries.
Kaufman, Z.S. 1968. The post embryonic period of development of some white sea starfish. *Doklady Akademii Nauk SSSR, Seriya Biologiya* 181:1009-12.
Kent, B.W. 1981. Behavior of the gastropod *Amphissa columbiana* (Prosobranchia: Columbellidae). *The Veliger* 23:275-76.
Khotimchenko, Y.S., I.I. Deridovich and E.A. Zalutskaya. 1985. Spectrofluorometric determination of indolylalkylamines in gonads of echinoderms. *Comparative Biochemical Physiology and Comparative Physiology* 81:457-59.
———. 1987. Identification of biogenic monoamines in sex glands of echinoderms and bivalve molluscs. *Soviet Journal of Marine Biology* 12:371-76.
Kishimoto, T., T.G. Clark, H.K. Kondo, H. Shirai and H. Kanatani. 1982. Inhibition of starfish oocyte maturation by some inhibitors of proteolytic enzymes. *Gamete Research* 5:11-18.
Kjerskog-Agersborg, H.P. 1918. Bilateral tendencies and habits in the twenty-rayed starfish *Pycnopodia helianthoides* (Stimpson). *Biological Bulletin* (Woods Hole, Mass.) 35:232-53.
Komatsu, M. 1986. Notes on the minute spicules of the sea star *Ctenodiscus crispatus*. *Proceedings of the Japanese Society of Sytematic Zoology* 33:45-50.

Kovesdi, I., and M.J. Smith. 1982. Sequence complexity in the maternal RNA of the starfish, *Pisaster ochraceus* (Brandt). *Developmental Biology* 89:56-63.

———. 1985. Actin gene number in the sea star *Pisaster ochraceus*. *Canadian Journal of Biochemistry and Cell Biology* 63:1145-51.

Kozloff, E.N. 1983. *Seashore Life of the Northern Pacific Coast*. Seattle: University of Washington Press.

———. 1987. *Marine invertebrates of the Pacific Northwest*. Seattle and London: University of Washington Press.

Kumé, M., and K. Dan. 1968. *Invertebrate Embryology*. Washington, D.C.: National Library of Medicine.

Kunz, C., and V.M. Conner. 1986. Roles of the home scar of *Collisella scabra* (Gould). *The Veliger* 29:25-30.

Kwast, K.E., D.W. Foltz and W.B. Stickle. 1990. Population genetics and systematics of *Leptasterias hexactis* (Echinodermata: Asteroidea) species complex. *Marine Biology* (Berlin) 105:477-89.

Lacalli, T.C. 1993. Ciliary bands in echinoderm larvae: evidence for structural homologies and a common plan. *Acta Zoologica* (Copenhagen) 74:127-33.

Lacalli, T.C., and J.E. West. 1993. A distinctive nerve cell type common to diverse deuterostome larvae: comparative data from echinoderms, hemichordates and Amphioxus. *Acta Zoologica* (Copenhagen) 74:1-8.

Lacalli, T.C., T.H.J. Gilmour and J.E. West. 1990. Ciliary band innervation in the bipinnaria larva of *Pisaster ochraceus*. *Philosophical Transactions of the Royal Society of London, Series B: Biological Sciences* 330:371-90.

Lafay, B., A.B. Smith and R. Christen. 1995. A combined morphological and molecular approach to the phylogeny of asteroids (Asteroidea: Echinodermata). *Systematic Biology* 44:190-208.

Lambert, P. 1978a. British Columbia marine faunistic survey report: Asteroids from the Northeast Pacific. *Fisheries and Marine Service Technical Report* No. 773:1-23.

———. 1978b. New geographic and bathymetric records for some north-east Pacific asteroids (Echinodermata: Asteroidea). *Syesis* 11:61-64.

———. 1981a. The Seastars of British Columbia. *British Columbia Provincial Museum Handbook* No. 39:1-153.

———. 1981b. The Sunflower Star. *Discovery: News and events from the Royal British Columbia Museum* 10:.

———. 1988. The sunflower star. *Royal British Columbia Museum Notes* No. 20:1-2.

———. 1994. In the clutches of a brooder! *The Victoria Naturalist* 51:5-7.

———. 1998. Stargazing in Alaska. *Discovery: News and events from the Royal British Columbia Museum* 26:4-5.

———. 1999. Range extensions for some Pacific coast sea stars (Echinodermata: Asteroidea). *Canadian Field Naturalist* 113(4):667-9.

Lansman, J.B. 1983a. Components of the starfish fertilization potential: role

of calcium and calcium dependent inward current neurology. *Neurobiology* 5:233-46.
———. 1983b. Voltage clamp study of fertilization in the egg of a starfish. *Dissertation Abstracts International [Section] B:* 43:2087.
———. 1983c. Voltage clamp study of the conductance activated at fertilization in the starfish egg. *Journal of Physiology* (London) 345:353-72.
———. 1987. Calcium current and calcium-activated inward current in the oocyte of the starfish *Leptasterias hexactis*. *Journal of Physiology* (London) 390:397-413.
Lawrence, J. 1987. *A functional biology of echinoderms*. London and Sydney: Croom Helm.
Lawrence, J.M. 1991. A chemical alarm response in *Pycnopodia helianthoides* (Echinodermata: Asteroidea). *Marine Behaviour and Physiology* 19:39-44.
LeClair, E.E. 1993. Effects of anatomy and environment on the relative preservability of asteroids: a biomechanical comparison. *Palaios* 8:233-43.
Legault, C., and J.H. Himmelman. 1993. Relation between escape behaviour of benthic marine invertebrates and the risk of predation. *Journal of Experimental Marine Biology and Ecology* 170:55-74.
Leighton, B.J., J.D.G. Boom, C. Bouland, E.B. Hartwick and M.J. Smith. 1991. Castration and mortality in *Pisaster ochraceus* parasitized by *Orchitophrya stellarum* (Ciliophora). *Diseases of Aquatic Organisms* 10:71-73.
Levin, A.V., E.V. Levina and V.S. Levin. 1984. The response of the asteroids *Asterias amurensis* and *Distolasterias nipon* to homogenates and chemical substances from Far Eastern starfishes. *Biologiya Morya (Vladivostok)* 1984:40-45.
Lissner, A., and D. Hart. 1996. Class Asteroidea. Pp. 97-112. In *Taxonomic atlas of the benthic fauna of the Santa Maria Basin and western Santa Barbara Channel*, vol. 14, edited by J.A. Blake, P.H. Scott and A. Lissner. Santa Barbara, California: Sant Barbara Museum of Natural History.
Littlewood, D.T.J., A.B. Smith, K.A. Clough and R.H. Emson. 1997. The interrelationships of the echinoderm classes: morphological and molecular evidence. *Biological Journal of the Linnean Society* 61:409-38.
Lopo, A.C., and V.D. Vacquier. 1980. Sperm-specific surface antigenicity common to seven animal phyla. *Nature* (London) 288:397-99.
MacGinitie, G.E., and N. MacGinitie. 1968. *Natural History of Marine Animals*. 2nd edition. Toronto, Ontario: McGraw-Hill Book Co.
Madsen, F.J. 1987. The *Henricia sanguinolenta* complex (Echinodermata: Asteroidea) of the Norwegian Sea and adjacent waters. A re-evaluation, with notes on related species. *Steenstrupia* 13:201-68.
Maluf, L.Y. 1988. Composition and distribution of the central eastern Pacific Echinoderms. *Natural History Museum of Los Angeles, Technical Reports* No. 2:1-242.
Manchenko, G.P. 1987. Electrophoretic estimation of the level of intraspecific genetic variability in sea-stars from the sea of Japan. *Soviet Journal of Marine Biology* 12:364-71.

Margolin, A.S. 1976. Swimming of the sea cucumber *Parastichopus californicus* (Stimpson) in response to sea stars. *Ophelia* 15:105-14.
Mauzey, K.P., C. Birkeland and P.K. Dayton. 1968. Feeding behaviour of asteroids and escape responses of their prey in the Puget Sound region. *Ecology* 49:603-19.
McCain, J.C. 1969. A new species of caprellid (Crustacea: Amphipoda) from Oregon. *Proceedings of the Biological Society of Washington* 82:507-10.
McClary, D.J., and P.V. Mladenov. 1988. Brood and broadcast: A novel mode of reproduction in the sea star *Pteraster militaris*. Pp. 163-68. In *Echinoderm Biology*, edited by R.D. Burke, P.V. Mladenov, P. Lambert and R.L. Parsley. Rotterdam: A.A. Balkema.
———. 1989. Reproductive pattern in the brooding and broadcasting sea star *Pteraster militaris*. *Marine Biology* (Berlin) 103:531-40.
———. 1990. Brooding biology of the sea star *Pteraster militaris* (O.F. Muller): energetic and histological evidence for nutrient translocation to brooded juveniles. *Journal of Experimental Marine Biology and Ecology* 142:183-99.
McClintock, J.B. 1989. The biochemical and energetic composition of somatic tissues during growth in the sea star, *Pycnopodia helianthoides* (Echinodermata, Asteroidea). *Comparative Biochemical Physiology and Comparative Physiology* 93:695-98,illustr.
McDaniel, N. 1971. The starfish *Solaster dawsoni* as a predator of asteroids. Honours BSc thesis. University of British Columbia, Vancouver.
———. 1977. The sun star - colourful creatures of the Pacific. *Diver and underwater adventure* 3:34-35.
McDaniel, N.G., C.D. Levings, D. Goyette and D. Brothers. 1978. Otter trawl catches at disrupted and intact habitats in Howe Sound, Jervis Inlet, and Bute Inlet, B.C., August 1967 to December 1977. *Fisheries and Marine Service Data Report* 92:1-74.
McEdward, L.R. 1992. Morphology and development of a unique type of pelagic larva in the starfish *Pteraster tesselatus* (Echinodermata: Asteroidea). *Biological Bulletin* (Woods Hole, Mass.) 182:177-87.
———. 1995. Evolution of pelagic direct development in the starfish *Pteraster tesselatus* (Asteroidea: Velatida). *Biological Journal of the Linnean Society* 54:299-327.
McEdward, L.R., and S.F. Carson. 1987. Variation in egg organic content and its relationship with egg size in the starfish, *Solaster stimpsoni*. *Marine Ecology Progress Series* 37:159-69.
McEdward, L.R., and F.-S. Chia. 1991. Size and energy content of eggs from echinoderms with pelagic lecithotrophic development. *Journal of Experimental Marine Biology and Ecology* 147:95-102.
McEdward, L.R., and L.K. Coulter. 1987. Egg volume and energetic content are not correlated among sibling offspring of starfish: implications for life history theory. *Evolution* 41:914-17.

Meijer, L., and P. Guerrier. 1981. Calmodulin in starfish oocytes. 1 Calmodulin antagonists inhibit meiosis reinitiation. *Developmental Biology* 88:318-24.

Meijer, L.M., and R.W. Wallace. 1980. Calmodulin in starfish oocytes. Pp. 385. In *Echinoderms - present and past. Proceedings of the European Colloquium on Echinoderms, Brussels, September 1979*, edited by M. Jangoux. Rotterdam, Netherlands: A.A. Balkema, P.O. Box 1675.

Meijer, L., and P. Zarutskie. 1987. Starfish oocyte maturation: 1 methyl-adenine triggers a drop of CAMP concentration related to the hormone-dependent period. *Developmental Biology* 121:306-15.

Meijer, L., S.L. Pelech and E.G. Krebs. 1987. Differential regulation of histone H1 and Ribosomal S6 kinases during sea star oocyte maturation. *Biochemistry* 26:7968-74.

Menge, B.A. 1972. Foraging strategy of a starfish in relation to actual prey availability and environmental predictability. *Ecological Monographs* 42:25-50.

———. 1975. Brood or broadcast? The adaptive significance of different reproductive strategies in the two intertidal sea stars *Leptasterias hexactis* and *Pisaster ochraceus*. *Marine Biology* (Berlin) 31:87-100.

Menge, B.A., E.L. Berlow, C.A. Blanchette, S.A. Navarrete and A.B. Yamada. 1994. Interaction strength in a rocky intertidal habitat. *Ecological Monographs* 64:249-85.

Menge, J.L., and B.A. Menge. 1974. Role of resource allocation, aggression and spatial heterogeneity in coexistence of two competing intertidal starfish. *Ecological Monographs* 44:189-209.

Miller, R.L. 1989. Evidence for the presence of sexual phermones in free spawning starfish. *Journal of Experimental Marine Biology and Ecology* 130:205-11.

Mladenov, P.V., B. Bisgrove, S. Asotra and R.D. Burke. 1989a. Mechanisms of arm-tip regeneration in the sea star, *Leptasterias hexactis*. *Roux's Archives of Developmental Biology* 198:19-28.

Mladenov, P.V., S. Igdoura, S. Asotra and R.D. Burke. 1989b. Purification and partial characterization of an autonomy-promoting factor from the sea star *Pycnopodia helianthoides*. *Biological Bulletin* (Woods Hole, Mass.) 176:169-75.

Moitoza, D.J., and D.W. Phillips. 1979. Prey defence, predator preference, and non-random diet: the interactions between *Pycnopodia helianthoides* and two species of sea urchins. *Marine Biology* (Berlin) 53:299-304.

Moody, W.J. 1985. The development of calcium and potassium currents during oogenesis in the starfish, *Leptasterias hexactis*. *Developmental Biology* 112:405-13.

Moody, W.J., and M.M. Bosma. 1985. Hormone-induced loss of surface membrane during maturation of starfish oocytes: differential effects on potassium and calcium channels. *Developmental Biology* 112:396-404.

Moody, W.J., and S. Hagiwara. 1982. Block of inward rectification by intracellular M+ in immature oocytes of the starfish, *Mediaster aequalis*. *Journal of General Physiology* 79:115-30.

Moody, W.J., and J.B. Lansman. 1983. Developmental regulation of Ca2+ and K+ currents during hormone-induced maturation of starfish oocytes. *Proceedings of the National Academy of Sciences of the United States of America (Biological science)* 80:3096-3100.

Morris, R.H., D.P. Abbott and E.C. Haderlie. 1980. *Intertidal Invertebrates of California*. Stanford, California: Stanford University Press.

Mortensen, T. 1921. *Studies of larval forms of echinoderms*. Copenhagen: Carlsberg Fund, G.E.C. Gad.

———. 1927. *Handbook of the echinoderms of the British Isles*. Oxford: Oxford University Press.

Moss, C., R.D. Burke and M.C. Thorndyke. 1994. Immunocytochemical localization of the neuropeptide S1 and serotonin in larvae of the starfish *Pisaster ochraceus* and *Asterias rubens*. *Journal of the Marine Biological Association of the United Kingdom* 74:61-71.

Motokawa, T. 1988. Catch connective tissue: A key character for echinoderm's success. Pp. 39-54. In *Echinoderm Biology*, edited by R.D. Burke, P.V. Mladenov, P. Lambert and R.L. Parsley. Rotterdam: A. A. Balkema.

Müller, J., and F.H. Troschel. 1840. Untitled. *Monatsberichte Preuss. Akademie Wissenschaften* 1840:100-106.

———. 1842. *System der Asteriden. 1 Asteriae. 2 Ophiuridae.* Braunschweig.

Nance, J.M., and L.F. Braithwaite. 1979. The function of mucous secretions in the cushion star *Pteraster tesselatus* Ives. *Journal of Experimental Marine Biology and Ecology* 40:259-66.

———. 1981. Respiratory water flow and production of mucus in the cushion star, *Pteraster tesselatus* Ives (Echinodermata: Asteroidea). *Journal of Experimental Marine Biology and Ecology* 50:21-31.

Obrietan, K., M. Drinkwine and D.C. Williams. 1991. Amylase, cellulase and protease activities in surface and gut tissues of *Dendraster excentricus*, *Pisaster ochraceus* and *Strongylocentrotus droebachiensis* (Echinodermata). *Marine Biology* (Berlin) 109:53-57.

O'Clair, C.E., and S.D. Rice. 1985. Depression of feeding and growth rates of the seastar *Evasterias troschelii* during long-term exposure to the water soluble fraction of crude oil. *Marine Biology* (Berlin) 84:331-40.

O'Clair, R.M., and C.E. O'Clair. 1998. *Southeast Alaska's rocky shores: Animals*. Auke Bay, Alaska: Plant Press.

Oguro, C. 1989. *Poraniopsis inflata* (Fisher) from Toyama Bay (Echinodermata: Asteroidea). *Bulletin of the Biogeographical Society of Japan* 44:49-50.

Otto, J.J., and T.E. Schroeder. 1984. Microtubule arrays in the cortex and near the germinal vesicle of immature starfish oocytes. *Developmental Biology* 101:274-81.

Paine, R.T. 1984. Ecological determinism in the competition for space. *Ecology* 65:1339-48.

Paine, R.T., J.C. Castillo and J. Cancino. 1985. Perturbation and recovery patterns of starfish-dominated intertidal assemblages in Chile, New Zealand, and Washington State. *American Naturalist* 125:679-91.

Palmer, A.R., J. Szymanska and L. Thomas. 1982. Prolonged withdrawal: a possible predator evasion behaviour in *Balanus glandula* (Crustacea: Cirripedia). *Marine Biology* (Berlin) 67:51-55.

Palumbi, S.R., and L.A. Freed. 1988. Agonistic interactions in a keystone predatory starfish. *Ecology* 69:1624-27.

Pathirana, C., and R.J. Andersen. 1986. Imbricatine, an unusual benzyltetrahydroisoquinoline alkaloid isolated from the starfish *Dermasterias imbricata*. *Journal of the American Chemical Society* 108:8288-89.

Patterson, M.J., J. Bland and E.W. Lindgren. 1978. Physiological response of symbiotic polychaetes to host saponins. *Journal of Experimental Marine Biology and Ecology* 33:51-56.

Patton, M.L., S.T. Brown, R.F. Harman and R.S. Grove. 1991. Effect of the anemone *Corynactis californica* on subtidal predation by sea stars in the southern California Bight. *Bulletin of Marine Science* 48:623-34.

Paul, A.J., and H.M. Feder. 1975. The food of the sea star *Pycnopodia helianthoides* (Brandt) in Prince William Sound, Alaska. *Ophelia* 14:15-22.

Pearse, J.S., and K.A. Beauchamp. 1986. Photoperiodic regulation of feeding and reproduction in a brooding sea star from central California. *International Journal of Invertebrate Reproduction and Development* 9:289-97.

Pearse, J.S., and D.J. Eernisse. 1982. Photoperiodic regulation of gametogenesis and gonadal growth in the seastar *Pisaster ochraceus*. *Marine Biology* (Berlin) 67:121-25.

Pearse, J.S., and A.H. Hines. 1987. Long-term population dynamics of sea urchins in a central California kelp forest: rare recruitment and rapid decline. *Marine Ecology Progress Series* 39:275-83.

Pearse, J.S., D.J. McClary, M.A. Sewell, W.C. Austin, A. Perez-Ruzafa and M. Byrne. 1988. Simultaneous spawning of six species of Echinoderms in Barkley Sound, British Columbia. *Invertebrate Reproduction and Development* 14:279-88.

Pelech, S. 1995. The profitable application of starfish oocytes for cancer research. *Bulletin of the Aquaculture Association of Canada* 95:9-13.

Pelech, S.L., L. Meijer and E.G. Krebs. 1987. Characterization of maturation-activated histone H1 and ribosomal S6 kinases in sea star oocytes. *Biochemistry* 26:7960-68.

Pelech, S.L., R.M. Tombes, L. Meijer and E.G. Krebs. 1988. Activation of myelin basic protein kinases during echinoderm oocyte maturation and fertilization. *Developmental Biology* 130:28-36.

Phillips, D.W. 1975. Distance chemoreception-triggered avoidance behavior of the limpets *Acmaea (Collisella) limatula* and *Acmaea (Notoacmaea) scutum* to the predatory starfish *Pisaster ochraceus*. *Journal of Experimental Zoology* 191:199-210.

Phillips, D.W. 1978. Chemical mediation of invertebrate defensive behaviors and the ability to distinguish between foraging and inactive predators. *Marine Biology* (Berlin) 49:237-43.
Polls, I., and J. Gonor. 1975. Behavioural aspects of righting in two asteroids from the Pacific Coast of North America. *Biological Bulletin* (Woods Hole, Mass.) 148:68-84.
Punnett, T., R.L. Miller and B.H. Yoo. 1992. Partial purification and some chemical properties of the sperm chemoattractant from the forcipulate starfish *Pycnopodia helianthoides* (Brandt, 1835). *Journal of Experimental Zoology* 262:87-96.
Quayle, D.B. 1954. Growth of the purple seastar. *Oyster Bulletin of the B.C. Department of Fisheries* 5:11-13.
Rasmussen, B.N. 1965. On the taxonomy and biology of the North Atlantic species of the asteroid genus *Henricia* Grey. *Meddelelser fra Danmarks Fiskeri-og Havundersogelser* 4:157-213.
Ray, S., D.W. McLeese, L.E. Burridge and B.A. Waiwood. 1980. Distribution of cadmium in marine biota in the vicinity of Belledune. *Canadian Technical Report of Fisheries and Aquatic Sciences* 963:11-34.
Reimer, C.L., and B.J. Crawford. 1990. Lectin histochemistry of the hyaline layer in the asteroid, *Pisaster ochraceus*. *Journal of Morphology* 203:361-75.
Reimer, C.L., B.J. Crawford and T.J. Crawford. 1992. Basement membrane lectin binding sites are decreased in the esophageal endoderm during the arrival of presumptive muscle mesenchyme in the developing asteroid *Pisaster ochraceus*. *Journal of Morphology* 212:291-303.
Rice, S.H. 1985. An anti-predator chemical defence of the marine pulmonate gastropod *Trimusculus reticulatus* (Sowerby). *Journal of Experimental Marine Biology and Ecology* 93:83-89.
Ricketts, E.F., J. Calvin and J.W. Hedgpeth. 1985. *Between Pacific Tides*. 5th edition. Stanford, California: Stanford University Press.
Robilliard, G.A. 1971. Feeding behaviour and prey capture in an asteroid, *Stylasterias forreri*. *Syesis* 4:191-95.
Rodenhouse, I.Z., and J.E. Guberlet. 1946. The morphology and behaviour of the cushion star *Pteraster tesselatus* Ives. *University of Washington Publications in Biology* 12:21-48, 4 plates.
Roller, R.A. 1988. Salinity effects on the development and larval tolerance of five species of echinoderms. *Dissertation Abstracts [Section] B: Sciences and Engineering* 48:2576-77.
Roller, R.A., and W.B. Stickle. 1985. Effects of salinity on larval tolerance and early developmental rates of four species of echinoderms. *Canadian Journal of Zoology* 63:1531-38.
Rosenberg, M.P. 1979. Relationship of changes in the cortical layers to the resumption of meiosis in starfish oocytes, hormonal and cation requirements. *Dissertation Abstracts International [Section] B:* 40:40.
Rosenberg, M.P., and H.H. Lee. 1981. The roles of Ca and Mg in starfish oocyte maturation induced by 1-methyladenine. *Journal of Experimental Zoology* 217:389-97.

Rowe, F.W.E., A.N. Baker and H.E.S. Clark. 1988. The morphology, development and taxonomic status of Xyloplax Baker, Rowe and Clark (1986) (Echinodermata: Concentricycloidea), with the description of a new species. *Proceedings of the Royal Society of London, Series B* 233:431-59.

Rowe, F.W.E., J.M. Healy and D.T. Anderson. 1994. Concentricycloidea. Pp. 150-67. In *Microscopic anatomy of invertebrates,* vol. 14, edited by F.W. Harrison and F.S. Chia. Toronto: Wiley-Liss.

Rumrill, S.S. 1989. Population size-structure, juvenile growth, and breeding periodicity of the sea star *Asterina miniata* in Barkley Sound, British Columbia. *Marine Ecology Progress Series* 56:37-47.

Sanghera, J.S., H.B. Paddon, S.A. Bader and S.L. Pelech. 1990. Purification and characterization of a maturation-activated myelin basic protein kinase from sea star oocytes. *Journal of Biological Chemistry* 265:52-57.

Sanghera, J.S., H.B. Paddon and S.L. Pelech. 1991. Role of protein phosphorylation in the maturation-induced activation of a myelin basic protein kinase from sea star oocytes. *Journal of Biological Chemistry* 266:6700-6707.

Schiel, D.R., and B.C. Welden. 1987. Responses to predators of cultured and wild red abalone, *Haliotis rufescens,* in laboratory experiments. *Aquaculture* 60:173-88.

Schroeder, T.E., and S.A. Stricker. 1983. Morphological changes during maturation in starfish oocytes: surface ultrastructure and cortical actin. *Developmental Biology* 98:373-84.

Schroeter, S.C., J. Dixon and J. Kastendiek. 1983. Effects of the starfish *Patiria miniata* on the distribution of the sea urchin *Lytechinus anamesus* in a southern Californian kelp forest. *Oecologia* 56:141-47.

Sewell, M.A., and J.C. Watson. 1993. A "source" for asteroid larvae?: recruitment of *Pisaster ochraceus, Pycnopodia helianthoides* and *Dermasterias imbricata* in Nootka Sound, British Columbia. *Marine Biology* (Berlin) 117:387-98.

Sheild, C.J., and J.D. Witman. 1993. The impact of *Henricia sanguinolenta* (O.F. Muller) (Echinodermata: Asteroidea) predation on the finger sponges, *Isodictya* spp. *Journal of Experimental Marine Biology and Ecology* 166:107-33.

Shick, J.M. 1976. Physiological and behavioural responses to hypoxia and hydrogen sulfide in the infaunal asteroid *Ctenodiscus crispatus. Marine Biology* (Berlin) 37:279-89.

Shick, J.M., K.C. Edwards and J.H. Dearborn. 1981a. Physiological ecology of the deposit feeding seastar *Ctenodiscus crispatus*:ciliated surfaces and animal-sediment interactions. *Marine Ecology Progress Series* 5:165-84.

Shick, J.M., W.F. Taylor and A.N. Lamb. 1981b. Reproduction and genetic variation in the deposit feeding sea star *Ctenodiscus crispatus. Marine Biology* (Berlin) 63:51-66.

Shirley, T.C., and W.B. Stickle. 1982a. Metabolic accommodation to salinity acclimation by *Leptasterias hexactis*. Pp. 357. In *Echinoderms: Proceedings of the Third International Echinoderm Conference, Tampa Bay*, edited by J.M. Lawrence. Rotterdam: A.A. Balkema.

Shirley, T.C., and W.B. Stickle. 1982b. Respones of *Leptasterias hexactis* (Echinodermata: Asteroidea) to low salinity. 1. Survival, activity, feeding, growth and absorption efficiency. *Marine Biology* (Berlin) 69:147-54.

———. 1982c. Responses of *Leptasterias hexactis* (Echinodermata: Asteroidea) to low salinity. 2. Nitrogen metabolism, respiration, and energy budget. *Marine Biology* (Berlin) 69:155-63.

Shivji, M., D. Parker, B. Hartwick, M.J. Smith and N.A. Sloan. 1983. Feeding and distribution study of the sunflower sea star *Pycnopodia helianthoides* (Brandt 1885). *Pacific Science* 37:133-40.

Simoncini, L., and W.J. Moody. 1990. Changes in voltage-dependent currents and membrane area during maturation of starfish oocytes: species differences and similarities. *Developmental Biology* 138:194-201.

Sladen, W.P. 1889. Report on the Asteroidea. *Report of the Scientific Results of the Voyage of HMS* Challenger *1873-76. Zoology* 30:1-893.

Sloan, N.A. 1977. An experimental study of the predatory and social behaviour of *Crossaster papposus* (L.). Ph.D. thesis. University of London.

———. 1979. Starfish encounters: an experimental study of its advantages. *Experientia* 35:1314-15.

———. 1980. The arm curling and terminal tube foot responses of the Asteroid *Crossaster papposus* (L.). *Journal of Natural History* 14:469-82.

———. 1984. Interference and Aggregation: close encounters of the starfish kind. *Ophelia* 23:23-31.

Sloan, N.A., and S.M. Northway. 1982. Chemoreception by the Asteroid *Crossaster papposus* (L.). *Journal of Experimental Marine Biology and Ecology* 61:85-98.

Sloan, N.A., and S.M.C. Robinson. 1983. Winter feeding by Asteroids on a subtidal sandbed in British Columbia. *Ophelia* 22:125-40.

Smirnov, A.V. 1994. Arctic echinoderms: Composition, distribution and history of the fauna. Pp. 135-43. In *Echinoderms through time*, edited by B. David, A. Guille, J.P. Feral and M. Roux. Rotterdam: A. A. Balkema.

Smith, L.S. 1961. Clam digging behaviour in the starfish, *Pisaster brevispinus* (Stimpson, 1857). *Behaviour* 18:148-53.

Smith, M.J. 1982. Molecular phylogeny of seastars. Pp. 209. In *Echinodermata: Proceedings of the Third International Echinoderm Conference, Tampa Bay*, edited by J.M. Lawrence. Rotterdam: A.A. Balkema.

Smith, M.J., D.K. Banfield, K. Doteval, S. Gorski and D.J. Kowbel. 1989. Gene arrangement in sea star mitochondrial DNA demonstrates a major inversion event during echinoderm evolution. *Gene* 76:181-85.

———. 1990. Nucleotide sequence of nine protein coding genes and 22 tRNAs in the mitochondrial DNA of the sea star *Pisaster ochraceus*. *Journal of Molecular Evolution* 31:195-204.

Smith, M.J., and R. Boal. 1978. DNA sequence organization in the common Pacific starfish, *Pisaster ochraceus*. *Canadian Journal of Biochemistry* 56:1048-54.
Smith, M.J., A. Lui, K.K. Gibson and J.K. Etzkorn. 1980. DNA sequence organization in the starfish *Dermasterias imbricata*. *Canadian Journal of Biochemistry* 58:352-60.
Smith, M.J., R. Nicholson, M. Stuerzl and A. Lui. 1982. Single copy homology in sea stars. *Journal of Molecular Evolution* 18:92-101.
Smith, R.I., and J.T. Carlton. 1975. *Light's Manual. Intertidal Invertebrates of the Central California Coast*. Berkeley, California: University of California Press.
Spies, R.B., and P.H. Davis. 1982. Toxicity of Santa Barbara seep oil to starfish embryos: part 3 - influence of parental exposure and the effects of other crude oils. *Marine Environmental Research* 6:3-11.
Steffen, W., and R.W. Linck. 1988. Evidence for tektins in centrioles and axonemal microtubules. *Proceedings of the National Academy of Sciences USA* 85:2643-47.
Stickle, W.B. 1985a. Effects of environmental factor gradients on scope for growth in several species of carnivorous marine invertebrates. *European Marine Biological Symposium* 18:601-16.
———. 1985b. Effects of stress on marine organisms. Effects of environmental factor gradients on scope for growth in several species of carnivorous marine invertebrates. Pp. 601-6. In *Marine biology of polar regions and effects of stress on marine organisms*, edited by J.S. Grey and M.E. Christiansen. Chichester, New York: John Wiley and Sons.
———. 1985c. Patterns of nitrogen excretion in seven species of asteroids. Pp. 557-62. In *Echinodermata: Proceedings of the Fifth International Echinoderm Conference, Galway/24-29 September 1984*, edited by B. Keegan F and B. O'Connor D.S. Rotterdam: A.A. Balkema.
Stickle, W.B., T.C. Shirley and T.D. Sabourin. 1982. Patterns of nitrogen excretion in four species of echinoderms as a function of salinity. Pp. 371-77. In *Echinoderms: Proceedings of the Third International Echinoderm Conference, Tampa Bay*, edited by J.M. Lawrence. Rotterdam: A.A. Balkema.
Stickle, W.B., D.W. Foltz, M. Katoh and H.L. Nguyen. 1992. Genetic structure and mode of reproduction in five species of sea stars (Echinodermata: Asteroidea) from the Alaskan coast. *Canadian Journal of Zoology* 70:1723-28.
Stimpson, W. 1862. On new genera and species of starfishes of the family Pycnopodidae (*Asteracanthion* Müll. and Trosch.). *Proceedings of the Boston Society of Natural History* 8:261-73.
Strathmann, M.F. 1987. *Reproduction and development of marine invertebrates of the northern Pacific coast: Data and methods for the study of eggs, embryos, and larvae*. Seattle and London: University of Washington Press.
Strathmann, R.R. 1971. The feeding behavior of planktotrophic echinoderm larvae: mechanisms, regulation, and rates of suspension feeding. *Journal of Experimental Marine Biology and Ecology* 6:109-60.

Stricker, S.A., and G. Schatten. 1991. The cytoskeleton and nuclear disassembly during germinal vesicle breakdown in starfish oocytes. *Development, Growth and Differentiation* 33:163-71.

Stricker, S.A., V.E. Centonze and R.F. Melendez. 1994a. Calcium dynamics during starfish oocyte maturation and fertilization. *Developmental Biology* 166:34-58.

Stricker, S.A., K.L. Conwell III, D.J. Rashid and A.M. Welford. 1994b. Nucleolar disassembly during starfish oocyte maturation. Pp. 29-43. In *Reproduction and development of marine invertebrates*, edited by W.H. Wilson Jr., S.A. Stricker and G.L. Shinn. Baltimore and London: Johns Hopkins University Press.

Taylor, R., and D.C. Williams. 1995. [Beta]-glucanase activity in the seastar *Pisaster ochraceus*. *Marine Biology* (Berlin) 123:735-40.

Tegner, M.J., and P.K. Dayton. 1987. El Nino effects on southern California kelp forest communities. *Advances in Ecological Research* 17:243-79.

Tezon, J., R. Miller and C.W. Bardin. 1984. Phospholipid methylation during chemotaxis of starfish spermatozoa. *Annals of the New York Academy of Sciences* 438:540-42.

Thomas, L.A., and C.O. Hermans. 1985. Adhesive interactions between the tube feet of a starfish, *Leptasterias hexactis*, and substrata. *Biological Bulletin* (Woods Hole, Mass.) 169:675-88.

Thomas, L.P. 1981. Living Color on the Rocky Shore. *Sea Frontiers* 27:150-51.

Turner, E., R. Klevit, L.J. Hager and B.M. Shapiro. 1987. Ovothiols, a family of redox-active mercaptohistidine compounds from marine invertebrate eggs. *Biochemistry* 26:4028-36.

Turner, R.L., and J.H. Dearborn. 1972. Skeletal morphology of the mud star, *Ctenodiscus crispatus* (Echinodermata: Asteroidea). *Journal of Morphology* 138:239-62.

Van Veldhuizen, H.D., and V.J. Oakes. 1981. Behavioural responses of seven species of asteroids to the asteroid predator, *Solaster dawsoni* (Responses of asteroids to the predator *Solaster dawsoni*). *Oecologia* 48:214-20.

Van Veldhuizen, H.D., and D.W. Phillips. 1978. Prey capture by *Pisaster brevispinus* (Asteroidea: Echinodermata) on soft substrate. *Marine Biology* (Berlin) 48:89-97.

Vermeij, G.J., R.B. Lowell, L.J. Walters and J.A. Marks. 1987. Good hosts and their guests: relations between trochid gastropods and the epizoic limpet *Crepidula adunca*. *The Nautilus* 101:69-74.

Verrill, A.E. 1880. Notice of the remarkable marine fauna occupying the outer banks off the southern coast of New England. *American Journal of Science* 20:390-403.

———. 1909. Description of new genera and species of starfishes from the North Pacific coast of America. *American Journal of Science* 28:59-70.

———. 1914. *Harrington Alaska Series. Vol.14. Monograph of the shallow-water starfishes of the North Pacific Coast from the Arctic Ocean to California*. New York: Smithsonian Institution.

Wagner, R.H., D.W. Phillips, J.D. Standing and C. Hand. 1979. Commensalism or mutualism: attraction of a sea star towards its symbiotic polychaete. *Journal of Experimental Marine Biology and Ecology* 39:205-10.

Walker, C.W. 1979. Ultrastructure of the somatic portion of the gonads in Asteroids, with emphasis on flagellated collar cells and nutrient transport. *Journal of Morphology* 162:127-62.

Ward, J.A. 1965a. An investigation of the swimming reaction of the anemone *Stomphia coccinea* 2. Histological location of a reacting substance in the asteroid *Dermasterias imbricata. Journal of Experimental Zoology* 158:365-72.

———. 1965b. An investigation on the swimming reaction of the anemone *Stomphia coccinea* 1. Partial isolation of a reacting substance from the asteroid *Dermasterias imbricata. Journal of Experimental Zoology* 158:357-64.

Watanabe, J.M. 1983. Anti-predator defenses of three kelp forest gastropods: contrasting adaptations of closely related prey species. *Journal of Experimental Marine Biology and Ecology* 71:257-70.

Webster, S.K. 1975. Oxygen consumption in echinoderms from several geographical locations, with particular reference to the Echinoidea. *Biological Bulletin* (Woods Hole, Mass.) 148:157-64.

Wobber, D.R. 1975. Agonism in Asteroids. *Biological Bulletin* (Woods Hole, Mass.) 148:483-96.

Wood, R.L., and M.J. Cavey. 1981. Ultrastructure of the coelomic lining of the podium of starfish *Stylasterias forreri. Cell and Tissue Research* 218:449-73.

Yayli, N. 1994. Isolation and characterisation of aromatic substituted olefinic compounds from the starfish *Pteraster militaris. Indian Journal of Chemistry [Section] B* June:556-61.

Young, C.M. 1985. Consequences of predation by the asteroid, *Evasterias troschelii* Stimpson on a soft sediment ascidian community. Pp. 577-83. In *Echinodermata: Proceedings of the fifth international echinoderm conference, Galway/24-29 September 1984,* edited by B. Keegan F and B. O'Connor D.S. Rotterdam: A.A. Balkema.

Young, C.M., P.G. Greenwood and C.J. Powell. 1986. The ecological role of defensive secretions in the intertidal pulmonate *Onchidella borealis. Biological Bulletin* (Woods Hole, Mass.) 171:391-404.

Zollo, F., E. Finamore, R. Riccio and L. Minale. 1989. Starfish saponins, part 37. Steroidal glycoside sulfates from starfishes of the genus *Pisaster. Journal of Natural Products (Lloydia)* 52:693-700.

ACKNOWLEDGEMENTS

I would like to thank the following people for assistance in putting together this edition. My field trip to southeast Alaska was made possible by a generous donation from Marna Disbrow. Molly Ahlgren and Walt Cunningham facilitated my field work in Sitka. Nora Foster helped me interpret the data at the University of Fairbanks Museum. In Juneau, Rita and Charles O'Clair provided gracious hospitality, Bruce Wing helped with the collections records and Bob Stone assisted me in diving. Special thanks to Chris Mah, California Academy of Sciences (CAS), for help with specimen records and for a thorough review of the manuscript. Bob van Syoc and Elizabeth Kools, CAS, also supplied a database of their museum records from Alaska. Frank Rowe provided many useful comments to improve the manuscript. Special thanks to Fu-Shiang Chia for permission to use his photograph of *Stylasterias* pedicellaria (I neglected to thank him in the first edition). Kathryn Oliver combed the library for all the references on sea stars since 1981, while she was a co-op student with me. Marilyn Lambert ably assisted me on the Alaska field work. For assistance with many dives in the Victoria region, I would like to thank my dive buddies Kelly Sendall, Gordon Green, Yves Parizeau, Nikki Wright and Michael Harvey. Susanne Beauchesne entered all the RBCM data on sea stars into a database.

From the first edition (1981):

I would like to express my thanks to Dr Arthur Fontaine, University of Victoria, who not only kindly relinquished his interest in preparing a handbook on sea stars but offered encouragement and constructive criticisms of the manuscript as well. My appreciation is also extended to Dr Bill Austin (Khoyatan Marine Laboratory), Dr Fu-Chiang Chia (University of Alberta), Maureen Downey (Smithsonian Institution) and Dr E. Kozloff (Friday Harbor Laboratories) for their reviews of the manuscript. Sincere thanks to my colleagues in the Aquatic Zoology Division of the British Columbia Provincial Museum [now the Royal British Columbia Museum], for assistance in field collections, processing of specimens and data and suggestions for improving the manuscript, and to Lynne Macdonald for typing the manuscript.

Neil McDaniel, formerly of the Pacific Environment Institute, kindly provided me with specific observations of the biology of sea stars. Thanks also to Dr Frank Bernard, Pacific Biological Station, for making available to me the asteroid collection of the Fisheries Research Board and assisting in clarifying the collection data; also to the National Museum of Canada for loan of specimens.

And finally, I am indebted to the authors of all the published works used in compiling the information in this handbook.

Sea Stars of British Columbia, Southeast Alaska and Puget Sound

Edited, designed and typeset by Gerry Truscott, RBCM.
The body type is Palatino 10/12; the headings are also in Palatino.
The figure captions and labels, and the running footers are in Optima.
Cover design by Chris Tyrrell, RBCM.

Front cover photograph by Philip Lambert.
Other photographs by Brent Cooke and Philip Lambert, RBCM, except:

- Figures 117 and 118: specimens prepared by T.H.J. Gilmour, University of Saskatchewan, and photographed by T.C. Lacalli. Printed by permission of Dr Gilmour.
- Figure 123: photographed by F.S. Chia, from Chia and Amerongen 1975. Reprinted by permission of the publisher.

Drawings by Gretchen Markle, except:

- Figure 1 (map) by Rick Pawlas.
- Figure 10 from Sloan and Robinson 1983. Reprinted by permission of the publisher.
- The Sea Daisy in figure 2 by Philip Lambert.

INDEX

Bold numbers indicate pages where families and species are described.